Integrated Water Resources Management in Latin America

It is now widely accepted that the world is likely to face a major water crisis unless the present management practices are significantly improved. Promoted extensively by donors and international institutions over the past 15 years, integrated water resources management (IWRM) has been assumed explicitly to be "the" solution for managing the limited water resources of the world. Hundreds of millions of dollars have now been spent in promoting IWRM in developing countries. However, no serious and objective analysis has ever been undertaken as to whether IWRM has made water management more efficient and equitable in any region of the world than otherwise may have been the case. This pioneering analysis indicates that IWRM has not only been unsuccessful in Latin America, but also is highly unlikely to succeed in the future. The reasons and constraints for this failure are outlined.

This book previously appeared as a special issue of the *International Journal of Water Resources Development*.

Asit K. Biswas is President of the Third World Centre for Water Management, Mexico, and Distinguished Visiting Professor at the Lee Kuan Yew School for Public Policy. He has advised over fifteen governments and his work has been translated into thirty-two languages.

Benedito P.F. Braga is Director of the National Water Authority, Brazil. He is also Vice President of the World Water Council and Past President of the International Water Resources Association.

Cecilia Tortajada is the Vice President of the Third World Centre for Water Management, Scientific Director of the International Centre for Water in Zaragoza, and Visiting Professor at the Lee Kuan Yew School for Public Policy, in Singapore.

Marco Palermo is the President of the Instituto Pro-Ambiente, an environmental non-governmental organization in Sao Paulo, Brazil.

Integrated Water Resources Management in Latin America

Edited by Asit K. Biswas, Benedito P.F. Braga, Cecilia Tortajada and Marco Palermo

INWAP

ANA
NATIONAL WATER AGENCY

Instituto
Pró-Ambiente

Routledge
Taylor & Francis Group

LONDON AND NEW YORK

First published 2009 by Routledge
2 Park Square, Milton Park, Abingdon, Oxon, OX14 4RN

Simultaneously published in the USA and Canada
by Routledge
270 Madison Avenue, New York, NY 10016

Routledge is an imprint of the Taylor & Francis Group, an informa business

Typeset in Times by Value Chain, India
Printed and bound in Great Britain by MPG Books Group, UK

British Library Cataloguing in Publication Data
A catalogue record for this book is available from the British Library

ISBN10: 0-415-48788-9
ISBN13: 978-0-415-48788-7

CONTENTS

Introduction: Integrated Water Resources Management in Latin America

During the past 15 years, much has been said and written about integrated water resources management (IWRM). In fact, one would be hard pressed to find a single national, regional or international organization in Latin America dealing with water that has not promoted IWRM in one way or another, and in one form or another, during the past decade. The emphasis given to IWRM may have varied from one institution to another, and at any one institution over time. However, it is fair to say that nearly all the water institutions of the region in recent years have considered IWRM at one time or another. The situation is somewhat similar in other regions of the world.

There is no doubt that IWRM received considerable attention during the International Conference on Water and the Environment held in Dublin in 1992. It received a further boost when the Global Water Partnership made IWRM a cornerstone of its technical programme, and the Summit for Sustainable Development, held in Johannesburg in 2002, recommended that every country should approve IWRM plans by 2005. Not surprisingly, with such major international forces backing this concept in recent years, many water institutions and professionals have sometimes considered IWRM to be the mantra that will solve all the world's water problems, irrespective of the type or scale of the problems, their geographical locations, or their governance environments.

Even though in recent years IWRM has been considered by most of the global water community as 'the solution', surprisingly few attempts have been made to analyze objectively and seriously the extent to which it has been applied successfully to improve water policies, programmes and projects at macro- or meso-scales. Furthermore, if and when the IWRM concept has been used, no serious attempt has been made to find out reliably what have been the improvements in terms of fulfilling the objectives of water management by using this concept which would not have happened without its use. Unfortunately, so strong has been the faith in IWRM in many quarters that these types of fundamental questions are not even being asked, let alone answered. Hundreds of millions of dollars are being spent each year to promote IWRM without seriously analyzing its implementation status in the real world, or determining its actual impacts in terms of better achieving the stipulated objectives of water resources management.

Because of the above and other unanswered associated questions, the Inter-American Development Bank (IDB), IDB-Netherlands Water Partnership Programme (INWAP), the National Water Agency of Brazil, Instituto Pro-Ambiente and the Third World Centre for Water Management organized a workshop in Rio de Janeiro, Brazil, to objectively review the status and extent of implementation of IWRM in the Latin American countries in order to determine whether the societal objectives of water management are being fulfilled better than before in a timely, cost-effective and socially acceptable manner. The meeting also considered the opportunities and constraints associated with the implementation of IWRM for large- to medium-size water projects. The participants were leading water experts from academic, national and international institutions, non-government organizations and the private sector, with special knowledge and expertise on water management in the Latin American region. Participation to the workshop was exclusively by invitation.

All the participants were invited in their personal capacities, and they did not represent or speak on behalf of their present or past institutions. The number of participants was restricted to 22 to ensure free, frank and in-depth discussions over three days. This book contains the papers that were prepared by some of the invited participants and formed the basis on which the initial discussions at the Rio de Janeiro Workshop took place.

There was unanimous agreement during the workshop that IWRM is a concept that is easy to talk about but has proved to be very hard to implement, even though IWRM is now a legal requirement in some Latin American countries, with penalties for non-compliance. Yet, there is no agreement at present as to how IWRM can be defined, or what is actually meant by IWRM, how it can be measured, or how the concept can be made operational. It would be fair to say that the attempts made thus far generally provide significant emphasis on integration, but not enough on how water resources can be optimally managed or how the expected societal objectives of water management can be better achieved. Nor has there been any serious discussion, let alone consensus, on what should be 'integrated' within the context of IWRM. An analysis of the existing literature on IWRM indicates that there are at least 42 different sets of issues that its proponents have recommended for integration, which is an impossible task because some of these issues are mutually exclusive. Furthermore, in terms of the knowledge and technology available at present, it is simply not feasible to integrate 42 different sets of issues, even if they were not mutually exclusive. Also, generally, the existing IWRM plans, or strategies, in Latin America have been strong on diagnosis, but weak on solutions and their eventual implementations. The situations are most likely to be similar in other regions of the world.

Because of the many very fundamental problems associated with the concept of IWRM and making it operational, the participants were hard pressed to identify even one good macro- or meso-scale IWRM project in Latin America which has been successfully operating for at least 10 years, and which has measurably improved the achievements of the goals and objectives of water projects which would not have occurred without the use of IWRM. This is in spite of the fact that the IWRM concept has been around for some half a century in one form or another. It was noted that at a scale of 1 to 100 (1 being no IWRM and 100 being full integration), it is difficult to find a single large water project anywhere in the region that can be given even a score of 30, based on medium- to long-term performance.

The papers prepared for the workshop and the resulting discussions clearly indicated that in terms of concept and implementation, IWRM at present has significantly more questions than answers. In addition, the focus of water planning and management needs to shift from 'means' such as IWRM to 'ends' such as poverty alleviation, regional development and environmental conservation. The focus needs to re-shift as to how these "ends" can be best reached for the specific locations in different parts of the world, including Latin America, which invariably have their own set of physical, social, economic, environmental, institutional and legal conditions.

The opportunities and constraints for implementing a 'means' such as IWRM vary with geographical locations. This is not a special situation exclusively for IWRM: it is equally valid for any other 'means'. If IWRM can deliver the 'end' best in a specific location, then this 'means' should be used. However, in a very heterogeneous world, one size does not fit all, and no one single 'means' is equally appropriate, or is the best solution, for all the countries of Latin America which have widely varying climatic, physical, social, economic and environmental boundary conditions. A logical and scientific approach requires that the 'means' or a solution that is most applicable and appropriate to reach the goals of a specific water management activity on a long-terms basis, in a specific location, should be selected. An a priori decision that IWRM is the best means to best fulfill the objectives of water resources management is unlikely to be universally valid, especially when existing conceptual and institutional constraints are considered. In a real world, based on past experience, the solution-in-search-of-a-problem approach has seldom proved to be a universal solution.

Equally, as the Latin American countries make economic progress, and knowledge and technology advance, the most appropriate and cost-effective solutions are likely to vary with time. For example, what may have been a good solution for Brazil or Mexico two decades ago, may not be the best one now. Equally, what may be a good solution at present for any particular country, is unlikely to be so in 20 years' time. In addition to the issue of scale noted earlier, time is also an important consideration, which regrettably has been almost totally neglected in the current IWRM discussions.

It is hoped that the papers of this book on the concept and implementation potentials of IWRM will stimulate regional as well as a global debate on how water can be better managed in an increasingly complex and interconnected world that is changing very rapidly, certainly at a much faster rate than ever before in history. Unfortunately, this type of debate has been conspicuous by it absence in an IWRM context.

The National Water Agency of Brazil and Instituto Pro-Ambiente have greatly served the water and development professions by organizing the workshop at Rio de Janeiro, which has raised many fundamental issues, considerations and constraints, in terms of the concept and implementation status of IWRM in a Latin American context. These issues, considerations and constraints now need to be seriously addressed and extensively debated to ensure a water-secure Latin America in the future. The workshop would not have been possible without the support of the Inter-American Development Bank and INWAP. However, the views expressed by all the authors in this publication are exclusively their own. They are not necessarily those of the sponsors or of the institutions with which the authors of the papers are affiliated.

Water is an important and critical development issue. It is also closely linked with other development sectors such as agriculture and energy, and social factors such as health, environment and quality of life. How it can best be managed in the future, with changing future conditions which are likely to be very different from the present or what has been witnessed in the past, needs significantly more attention than it is currently receiving. It is becoming increasingly apparent that tomorrow's water problems can no longer be optimally managed with today's approaches, yesterday's mind-sets and the day- before-yesterday's experience. Business-as-usual can no longer be the solution. We need to consider 'business-unusual' approaches.

Asit K. Biswas, President
Third World Centre for Water Management
Atizapan, Mexico

Integrated Water Resources Management: Is It Working?

ASIT K. BISWAS

Introduction

According to the Greek philosopher Pindar, the best of all things is water. This view is not surprising since the need for water, throughout human history, has always been appreciated. It is present everywhere, and without water, life, as it is known, will simply cease to exist. Water is constantly in motion, passing from one state to another and from one location to another. Whether the water is in motion, or stationary as it is in lakes, it invariably contains extraneous materials, some due to natural causes but others because of human activities. All these, plus natural variations in water availability, makes its rational planning and management a very complex and difficult task under the best of circumstances. Water may be everywhere, but its use has always been dictated by its availability in terms of quantity and quality.

Water problems of the world are neither homogenous, nor constant or consistent over time. They often vary very significantly from one region to another, sometimes even within a single country, from one season to another, and also from one year to another. Solutions to water problems depend not only on water availability, but also on many other factors, among which are the processes through which water is managed; competence and capacities of the institutions that manage them; prevailing socio-political conditions and expectations which affect water planning, development and management processes and practices; appropriateness and implementation statuses of the legal and regulatory frameworks; availability of investment funds as and when needed; climatic, social and environmental conditions of the countries concerned; levels of available and usable technology; national, regional and international attitudes and perceptions; modes of governance including issues like political interference, transparency, corruption, etc.; educational and development conditions; and quality, effectiveness and relevance of research that are being conducted to solve the national, sub-national and local water problems.

Water is a resource that is of direct interest to the society as a whole, as well as to most development-related public institutions at central, state and municipal levels, academia, private sector and non-governmental organizations (NGOs). Such widespread interest in water is not a unique situation, as many water professionals have often claimed: it is equally applicable to other important sectors like food, energy, the environment, health, communication or transportation. All these issues command high levels of social and political attention in all modern societies, although their relative importance may vary from one country to another, and also over time. In an increasingly interrelated and complex world, many issues are of pervasive interest for assuring good quality of life of the people. Water is one of these important intersectoral issues, but it is certainly *not* the only issue, or often the *most* important socio-political issue, irrespective of the views of many in the water profession. In recent years, it has become increasingly evident that the water problems of a country can no longer be resolved exclusively by the water professionals, and/or the water ministries, alone. The water problems are becoming increasingly more and more interconnected and intertwined with other development-related issues, and also with social, economic, environmental, legal and political considerations, at local and national levels, and sometimes even at regional and international levels. Many of the water problems have already become far too complex, interconnected and large to be handled by any one single institution, irrespective of the authority and resources given to it, technical expertise and management capacity available, level of political and public support, and all the good intentions (Biswas, 2001).

The current and the foreseeable trends indicate that water problems of the future will continue to become increasingly complex, and will become more and more interlinked with other development sectors such as agriculture, energy, industry, transportation and communication, and with social sectors such as education, the environment, health and rural or regional development (Asian Development Bank, 2007). The time is fast approaching when water can no longer be viewed in isolation by primarily one single institution, or any one group of professionals, without explicit and simultaneous consideration of other related sectors and issues that affect water management, and vice versa. In fact, it can be successfully argued that the time has already come when water policies and major water-related issues should be assessed, analysed, reviewed and resolved within an overall societal and development context, otherwise the main objectives of water management, such as improved standard and quality of life of the people, poverty alleviation, regional and equitable income distribution and environmental conservation, cannot be achieved. One of the main questions facing the water profession is how this challenge can be successfully answered in a socially acceptable and economically efficient manner.

During the past 15 years or so, and heavily promoted by the donors, the mantra has often been that integrated water resources management will solve the water problems everywhere in the world, in spite of the different physical, economic, social and environmental conditions of a very heterogeneous world, and irrespective of the rapidly increasing complexities of water management practices and processes. The present paper analyzes how realistic this widely promoted universal solution is to the water management problems of the world.

Integrated Water Resources Management: Background and Definition

During the early 1980s, a few members of the water profession started to realize that the overall global water situation was not as good as it appeared. This feeling intensified during the 1990s, when many more in the profession began to appreciate that the water problems had become multi-dimensional, multi-sectoral and multi-regional, and were enmeshed with multi-interests, multi-agendas and multi-causes, which could be resolved only through an appropriate multi-disciplinary, multi-institutional and multi-stakeholders coordination. However, at present the main question is not whether such a process is desirable, but rather can this be achieved in the real world in a timely, cost-effective and socially acceptable manner?

Faced with such unprecedented management complexities, many in the water profession started to look for a new paradigm for management, which would solve the existing and the foreseeable problems in different parts of the world. However, the solution that was selected and which became increasingly popular was not new. It was the rediscovery of a basically more than 60-year old concept, which could not be successfully implemented previously: integrated water resources management. Many who 'discovered' this concept were not even aware that the 'new' concept was in fact not at all new, but had been around for several decades, with a dubious implementation record, which had never been objectively, comprehensively and critically assessed.

Before the status of application of integrated water resources management can be discussed, an important and fundamental issue that needs to be first considered is what precisely is meant by this concept. A comprehensive and objective assessment of the recent writings of the individuals and the institutions that have vigorously championed integrated water resources management indicates that not only no one has a clear idea as to what exactly this concept means in operational terms, but also their views of it in terms of what it actually means and involves, vary very widely. It can even be argued that this very vagueness has contributed to the high popularity of the integrated water resources management concept since people could continue to do what they had done before, or are doing at present, but put these activities under an increasingly popular bandwagon for which considerable resources have been made available by the donors and international institutions.

The definition that is most often quoted at present is the one that was formulated by the Global Water Partnership (GWP, 2000), which started to champion integrated water resources management as a major component of its technical programme shortly after its inception. GWP defined it as:

> a process which promotes the coordinated development and management of water, land and related resources, in order to maximize the resultant economic and social welfare in an equitable manner without compromising the sustainability of vital ecosystems.

This definition, on a first reading, appears broad, all-encompassing and, perhaps even impressive, at least linguistically. However, such lofty phrases, when scrutinized carefully and objectively, have little practical resonance on the present, or on future water management practices. A serious and critical look at this amorphous definition may remind one of the immortal writings of William Shakespeare:

Polonius: What do you read, my lord?

Hamlet: Words, words, words.

Unfortunately, for a variety of reasons, a fundamental question that has never been asked, let alone answered, either by the GWP or the promoters of this paradigm who have uncritically accepted the GWP definition as the gospel, is that whether this well-intentioned and good-sounding definition has any practical value in terms of its application and implementation to improve existing water management, or is it just an aggregation of trendy words which collectively provides an amorphous definition which does not help water planners and managers very much in terms of the application of the concept to solve the real water-related problems that are being faced in different parts of the world.

Let us consider only some of the fundamental questions that the above definition raises in terms of its possible application in the real world, which have not been addressed to in any significant way thus far, either by GWP or by the proponents of IWRM. Among these questions are the following:

- 'Promotes': Who promotes this concept? Why should it be promoted, and through what processes? Can the promotion of an amorphous concept be enough to improve water management? What about its implementation?
- 'Land and related resources': What is meant by 'related resources'? Does it include agriculture, energy, minerals, fish, other aquatic resources, forests, the environment, etc.? Even if only land and agricultural resources are considered, the institutions responsible for water management have seldom any say, or authority, over them. Considering the intense inter-ministerial and intra-ministerial rivalries that have always been present in all countries, how can the use, development and management of land and agricultural resources be integrated with water, even if this was technically, administratively, knowledge-wise and managerially possible? Is this realistically feasible? If the boundaries of integration are further expanded, and issues such as the environment and ecosystems are considered, how can the water professionals and the relevant ministries handle such integration, which is often beyond their knowledge, expertise and/or legal and institutional control?

  Interestingly, but not surprisingly, the people who formulated this definition for the Global Water Partnership were all from the water profession: experts from 'land and related resources' were singularly conspicuous by their absence, as were from other resource-related professions. This raises one fundamental question, that is, what makes the water profession believe that they can superimpose their views on the other professions, who were not even consulted and on which they have only limited knowledge and expertise? Equally, why should the professionals from other professions accept the view of some people from the water profession? A cynic might even be excused for claiming that the water profession prefers to remain in water-tight compartments, but preach integration with other sectors without any consultations or discussions with the professionals of appropriate disciplines, sectors and institutions.

- 'Maximize': What specific parameters should be maximized? What process should be used to select these parameters adequately and reliably? Who will select these parameters: only water experts, as was the case for the formulation of the GWP definition, or should professionals from other sectors be involved? What criteria should be used to select the necessary parameters? What reliable methodology is available at present to maximize the selected parameters? Do such methodologies even exist at present? If not, can they be developed within a reasonable timeframe so that these can be used?

- 'Economic and social welfare': What exactly meant by economic and social welfare? Even the economists and the sociologists cannot agree as to what actually constitutes economic and social welfare, except in somewhat general and broad terms. How can the issues related to social and economic welfare be quantified? Can these be even quantified? Are water professionals capable of maximizing economic and social welfare in operational terms, a fact that has mostly eluded even the social scientists thus far? Is it possible that even the cause-and-effect relationships between water development and management and economic and social welfare can be established, let alone be maximized? Such functional relationships are mostly unknown at present. Even if they were known, which they are not, they are likely to be a site- or region-specific, and thus generalization simply will not be possible on a global scale, as is implied by the definition.

- 'Equitable': What is precisely meant by equitable? How will this be determined operationally? Who will decide what is equitable, for whom, and from what perspectives and under what conditions?

- 'Sustainability': What is meant by sustainability, which itself is as a vague word, and perhaps also as fashionable and trendy, as integrated? How can sustainability be defined and measured in operational terms?

- 'Vital ecosystems': What exactly constitutes vital ecosystems? How can 'vital' and 'non-vital' ecosystems be differentiated? Can such a differentiation even be made in conceptual terms, let alone in operational and implementation terms? What are the minimum boundary conditions that will ensure the 'sustainability' of the 'vital ecosystems', at least in terms of its linkage to water, irrespective of how sustainability itself is defined, or the issue of what constitutes vital ecosystems is resolved?

When all these uncertainties and unknowns are aggregated, the only objective and realistic conclusion that can be drawn is that even though on a first reading the definition formulated by the Global Water Partnership appears impressive, it has to be admitted by any objective person that it is simply unusable, or unimplementable, in operational terms. Not surprisingly, even though the rhetoric of integrated water resources management has been very strong at many international and national fora during the past decade, its actual use (irrespective of what it means) has been minimal, even indiscernible in the field (for an analysis of its actual use in south and southeast Asia, see Biswas *et al.*, 2004, and for Latin

America see Biswas *et al.*, 2008). In fact, it can even be successfully argued that it would not have made any perceptible difference in enhancing the efficiencies of macro- and meso-scale water policies, programmes and projects of the recent years, even if the concept of integrated water resources management had not been resurrected, reinvented and promoted vigorously by the various donors and international institutions in recent years.

No objective person will question that for all practical purposes, the definition that has been formulated by the Global Water Partnership is unusable and unimplementable. In addition, it is internally inconsistent. Furthermore, while the definition has effectively collated many of the recent trendy, fashionable and politically correct words, it does not provide any real guidance to the water professionals and policy makers as to how the concept can be operationalized to make the existing water planning, management and decision-making processes increasingly more and more rational and efficient so that the actual objectives of water management can be achieved.

What Issues Should Be Integrated?

Analyses of existing literature indicate that the authors concerned have considered different issues that need to be integrated under this concept. This is not surprising, since as noted earlier, there is simply no agreement in the profession as to what integrated water resources management means, and what it really entails.

The word 'integration' often has had very different connotations and interpretations depending on the author(s) and institutions concerned, and their interests. Depending upon the author(s) and/or institutions, integrated water resources management requires integration of:

- objectives which are not mutually exclusive (economic efficiency, regional income redistribution, environmental quality and social welfare);
- water supply and water demand;
- surface water and groundwater;
- water quantity and water quality;
- water and land-related issues;
- different types of water uses: domestic, industrial, agricultural, navigational, recreational, environmental and hydropower generation;
- rivers, aquifers, estuaries and coastal waters;
- water, the environment and ecosystems;
- water supply and wastewater collection, treatment and disposal;
- urban and rural water issues;
- irrigation and drainage;
- water and health;
- macro, meso and micro water projects and programmes;
- water-related institutions at national, regional, municipal and local levels;
- public and private sectors;
- government and NGOs;
- timing of water release from the reservoirs to meet domestic, industrial, agricultural, navigational, environmental and hydropower generation needs;
- all legal and regulatory frameworks relating to water, not only from the water

sector, but also from other sectors that have direct implications on the water sector;
- all economic instruments that can be used for water management;
- upstream and downstream issues and interests;
- interests of all different stakeholders;
- national, regional and international issues;
- water projects, programmes and policies;
- policies of all different sectors that have water-related implications, both in terms of quantity and quality, and also direct and indirect (sectors include agriculture, industry, energy, transportation, health, the environment, education, gender, etc.);
- intra-state, interstate and international rivers;
- bottom-up and top-down approaches;
- centralization and decentralization;
- national, state and municipal water activities;
- national and international water policies;
- timings of water release for municipal, hydropower, agricultural, navigational, recreational and environmental water uses;
- climatic, physical, biological, human and environmental impacts;
- all social groups, rich and poor;
- beneficiaries of the projects and those who pay the costs;
- service providers and beneficiaries;
- present and future generations;
- national needs and interests of donors;
- activities and interests of donors
- water pollution, air pollution and solid wastes disposal, especially in terms of their water linkages;
- various gender-related issues;
- present and future technologies;
- water development and regional development; and
- any number of formulations and combinations of the above.

The above list, which is by no means exhaustive, identifies at least 41 sets of issues which different authors and/or institutions consider to be the issues that should be integrated under the aegis of integrated water resources management. Even at a conceptual level, all, or even many of these 41 sets of issues that the proponents would like to be integrated, simply cannot be achieved. At our present state of knowledge, this simply cannot be done. Nor is it likely to be achieved in the foreseeable future.

These types of fundamental issues and constraints need to be discussed and resolved successfully before the concept of integrated water resources management can be considered to be an universal approach to improve water management, as has been promoted in recent years. It is highly unlikely that these issues and constraints can be resolved, or one solution can be found which can be implemented all over the world. These are totally unrealistic expectations.

Unfortunately, while much lip-service has been given to this concept in recent years, most of the published works on the subject are somewhat general, or a continuation of earlier 'business as usual' approaches, but with a trendier label of integrated water resources management. If integrated water resources management is ever to become successful approach to water management, national and international organizations will have to address many real and complex questions, which they have not done so far in any meaningful fashion, nor is there any indication whatsoever that they are likely to do so in the foreseeable future. Under these circumstances, and unless the current rhetoric can be translated effectively into operational reality, integrated water resources management will remain a fashionable and trendy concept for another few years, and then gradually fade away like many other similarly popular concepts of the earlier times. There are already some signs that this is already happening, since a few of its ardent past promoters have stopped promoting this concept.

Implementational Constraints

The definition of integrated water resources management is an important consideration. When the definitional problem can be successfully resolved in an operational manner, it may be possible to translate it into measurable criteria, which can then be used to appraise the degree to which the concept of integration has been implemented in a specific case, and also the overall relevance, usefulness and effectiveness of the concept in terms of improving practices and processes used for water management.

In addition, a fundamental question that has never been asked, let alone answered, or for which there is no clear-cut answer at the present state of knowledge, is what are the parameters that need to be monitored to indicate that a water resources system is functioning in an integrated manner, or a transition is about to occur from an integrated to an 'unintegrated' stage, or vice versa, or indeed even such a transition is occurring? In the absence of both an operational definition and measurable criteria, it is not possible to identify what actually constitutes an integrated water resources management system at present, or how water should be managed so that the system remains inherently integrated on a long-term basis.

Nor have the proponents of the concept given any serious thought to the data requirements for the application of this concept. Irrespective of all the intensive promotion of this paradigm, what type and extent of data are needed to implement this concept in the real world, assuming that somehow it can ever be implemented? Are such levels of data available even in developed countries, let alone in developing countries? This is an important topic that is considered in further detail in this issue of the journal by Rachael McDonnell (2008). In addition, the Asian Development Bank (2007) has raised the serious issue of paucity and reliability of data on all aspects of water-related issues in the Asian developing countries. The proponents of integrated water resources management concept have been conspicuous by their neglect of the data availability, reliability and accessibility issues.

There is no question that in the water area, integrated water resources management has become a powerful and all-embracing slogan during the past 15 years. This is in spite of the fact that operationally it has not been possible to identify a water management process at a macro- or meso-scale which can be planned and implemented in such a way that it becomes inherently integrated, however this may be defined, right from its initial planning stage and then to implementation and operational phases. For all practical purposes, most international institutions have endorsed this concept, either explicitly or implicitly, without seriously analysis of its usability and implementability. This is in spite of the facts that there is no agreement at present among the various international institutions that endorse it as to what exactly is meant by integrated water resources management, or whether this concept has improved water management practices anywhere in the world, which would not have occurred otherwise without the explicit use of this concept. Furthermore, in which countries, if any, this concept has been successfully implemented, and, if so, under what conditions, over what periods, and what have been its impacts (positive, negative and neutral) on human lives, the environment and other appropriate development indicators. Even the donors who have been promoting this concept vigorously will be hard-pressed to identify even one good case at successful implementation of integrated water resources management in their own countries. Not surprisingly, increasingly more and more national and international institutions and water professionals have started to question the relevance and the appropriateness of the implementation potential of integrated water resources management.

As noted earlier, this type of almost universal popularity of a vague, undefinable and unimplementable concept is not a new phenomenon in the area of natural resources management. It has happened many times earlier. For example, during the 20th century many popular concepts have come and gone, without leaving much of a footprint on how natural resources can be managed efficiently on a long-term basis. Such concepts generally became politically correct during the time of their popularity, and are widely embraced since they are vague enough for everyone to jump on the bandwagon and claim that they are following the latest development. In fact, it appears that the vagueness of a concept, to a significant extent, increases its popularity, since people can then continue to do the same old stuff (SOS) they were doing before, but can concurrently claim that they are au currant with the latest global thinking. This jumping on the bandwagon also increases, often very significantly, the potential of receiving funding support from the donors, and also other personal benefits.

The current popularity of the concept reminds one of another similar concept which received wide popular support in the United States during the early 20th century: conservation. Even President Roosevelt of the United States said at that time that: "Everyone is for conservation: no matter what it means!" (Biswas, 2001). The situation has been somewhat similar in recent years with integrated water resources management. To paraphrase, and perhaps update President Roosevelt, it can be said that "Everyone is for integrated water resources management: no matter what it means, no matter whether it can be implemented, or no matter whether it would actually improve water management processes". However, there is an important difference between the Conservation Movement witnessed during President Roosevelt's time and the current push by the donors for integrated water resources management. This is because information and communication revolution and globalization processes have ensured that the gospel of integrated water resources management could be spread quickly all over the world, and not mostly confined to one country, as was the case for the Conservation Movement earlier. Strong funding support and political push from the donors have further contributed to the increased global spread of integrated water resources management. These were not important factors for the Conservations Movement.

Is Integrated Water Management a New Concept?

Shortly after the Dublin Conference in 1992, and following the embracement by GWP of integrated water resources management as a main component of their programme, the concept gained traction from several international institutions during the 1990s, many of whom were not even aware that the concept had been around for more than half a century! Accordingly, and not surprisingly, the authors of Toolbox for IWRM for the Global Water Partnership claimed, totally erroneously, in 2003, that "IWRM draws its inspiration from the Dublin principles", being blissfully unaware of the longevity of this concept, or the fact that international institutions such as the United Nations were promoting this concept extensively during the 1950s, or that the United Nations Water Conference, held in Mar del Plata, Argentina, in March 1977 had more relevant statements on integrated water resources management (Biswas, 1978) than the Dublin Conference. In addition, the Mar del Plata Conference was an intergovernmental meeting, and its Action Plan (which included references to integrated water resources management and other appropriate means for water management) was endorsed by all the governments that were members of the United Nations in 1977. In contrast, the Dublin Conference of 1992 was a meeting of experts. Accordingly, its recommendations, whatever may be their values or relevance, were never approved by the global community of the governments, irrespective of the claims to the contrary of the individuals and institutions that were responsible for the organization of the Dublin Conference, many of whom subsequently became the major promoters of IWRM. Thus, to a significant extent, many of the post-Dublin proponents of integrated water resources management not only rediscovered the wheel, but also the wood with which the wheel was made of!

It should be noted that the Global Water Partnership spent very considerable resources in developing and promoting the so-called Toolbox for integrated water resources management. The examples provided in the Toolbox have never received objective scrutiny or serious peer-review, and no objective and independent evaluation was ever made to determine if the so-called 'tools' were actually used and resulted in improving water management measurably which would not have happened otherwise. Nor was the replicability potential of the various 'tools' was ever seriously considered or objectively assessed. Under these conditions, and, not surprisingly, the global interest in the Toolbox, for all practical purposes, has basically disappeared, irrespective of the fact that immense amount of resources and efforts were expended to develop and promote the Toolbox.

Other Considerations

Extensive and intensive analyses of integrated water resources management literature published during the past decade indicate three unwelcome developments. First, there is no clear understanding of what exactly integrated water resources management means. Accordingly, different people have interpreted this concept very differently, but under a very general catch-all concept of integrated water resources management. The absence of any usable and implementable definition has only compounded the vagueness of the concept, and has reduced its implementation potential to a minimum. Second, because of the recent popularity of the concept, many people and institutions have continued to do what they were doing in the past, but under the guise of integrated water resources management in order to attract additional funds, or to obtain greater national and international acceptance and visibility. Third, considerable efforts have been expended by the various donors to promote the concept extensively, but irrespective of their oft-repeated rhetoric, the results have been meagre.

An analysis of the recently published literature on only one of the definitional aspects of the concept, that is, what are the issues that should be integrated, under the IWRM level, indicates a very wide divergence of opinions. It should be noted that this refers only to what should be integrated, and *not* to other equally important fundamental issues such as how can these issues be integrated (even if they can actually be integrated since many of the issues are mutually exclusive), who will do the integration and why, what processes will be used for integration (do such processes currently exist?), or will the integration, if at all it can be done, produce the benefits that proponents have claimed. Regrettably, none of these questions have ever been asked seriously in the past and are not being asked now. Not surprisingly, at present there are no objective and definitive answers to such fundamental questions. Consequently, acceptance of the concept has been primarily a leap of faith, and not based on its scientific merit or technical strength.

Another very unwelcome development has been that the current high priests of integrated water resources management, for the most part, have refused to argue in public on the validity and applicability of the concept with those who have questioned it. Instead, a deliberate attempt has often been made to ostracise and denigrate the opponents of the concept, and, sometimes, attempts have even been made to cut off their funding sources through backdoor channels. Sadly, the proponents have made no attempt to win the intellectual and technical arguments behind integrated water resources management. Unfortunately, they have either forgotten or have found it convenient to forget, a fundamental principle of science and knowledge generation. Knowledge does not advance by consensus: if it did, we would still be living in the Dark Ages! (Biswas, 2006)

Popularity of the Concept

An important issue that needs to be asked is why an old concept suddenly became so popular in the 1990s, to the extent that some people and institutions even considered it to be the 'holy grail' of water management? There are many reasons for its sudden leap of popularity, and only some of the main reasons will be discussed herein.

Probably one of the two most important reasons for its current popularity is the simplicity of the concept: it is easy to understand at a conceptual level, at least at a first glance. In a world that operates on the principle of reductionism, integrated water resources management often gives a false feeling of using a comprehensive and holistic approach, which many people a priori assume will produce the best results, irrespective of its inherent shortcomings and numerous fundamental inconsistencies embedded in the concept. These constraints and complexities need to be objectively and comprehensively assessed.

The second reason for its popularity is because of the amount of funds the donors have pumped in promoting this concept. This enormous level of funding has been primarily responsible for the creation of a new and thriving industry on integrated water resources management. This development is, of course, not new. For example, as Hall (2003) has perceptibly noted:

> One needs to be realistic about how humans, universities and research institutions work. They are driven by egos and money. For example, when research on any issue starts getting hot, soon by land, sea and air, the field is invaded by researchers scrambling for a piece of action, pursuing their intellectual curiosity with all the decorum and dignity of the 19th century gentlemen geologists who pursued their curiosity about rumours of gold in California.

As long as the donors continue to pump money in promoting the concept, the bandwagon will keep rolling, until the countries whose water management were supposed to have improved by this old-wine-in-a-new-bottle concept realize that they are making no visible progress. Fortunately, there are now increasing signs that some donors are now carefully evaluating the validity and applicability of integrated water resources management as a universal solution, and some developing countries are assessing whether this concept, which they have made national policies at the urging of donors and international institutions, has produced the expected benefits. All these objective reassessments should be considered to be necessary developments.

Need for Reductionism

Historically, it was possible for a brilliant person to know nearly all there was to know until about the end of the 16th century. Versatile geniuses such as Aristotle, Theophrastus, Vitruvias, Isidore of Seville and Leonardo da Vinci could discuss most subjects authoritatively. Human knowledge, in terms of natural and social sciences, was at a stage where it was possible for a truly gifted person to master all the knowledge that was available during their lifetimes.

The situation started to change around the 17th century. By the early 18th century, tremendous advances in knowledge had made it impossible for anyone to be a universal encyclopaedist, and keep up with the constant generation of new knowledge. This realization was gradually reflected in the development of a new branch of knowledge, which initially became known as natural philosophy, and began to be distinguished increasingly from traditional philosophy, which was earlier considered to be the exclusive discipline for knowledge. The 19th century witnessed exponential advances in human knowledge and, with it, technological developments. It was no longer possible for anyone individual to master even natural philosophy completely. Thus, new disciplines began to emerge, which further fragmented the knowledge-base to manageable levels. Natural philosophy was subsequently subdivided, initially into physics and chemistry, and later to other additional disciplines such as life sciences and biological sciences.

The knowledge and information explosion of the 20th century further accelerated this reductionism trend. Disciplines became more and more fragmented. It became humanly impossible for anyone to know everything there is to know even in a much more restricted subject area such as water. Knowledge, communication and information revolution and increasing globalization witnessed towards the end of the 20th century further constrained the mastering of a person's disciplinary knowledge-base. With the frontiers of knowledge expanding continuously, it has become increasingly difficult for professionals to keep up with the advances even in their limited areas of interest, such as water.

As the world became increasingly complex and interrelated, the disciplinary knowledge-base of individuals started to reduce as well. People started to specialize in narrower and narrower subject areas. Concomitantly, managing human societies became increasingly complex, as a result of which new institutional machineries had to be created with increasingly narrower focuses. New institutions had to be created in areas that were part of broader groups earlier. For example, in 1972, when the United Nations Conference on the Human Environment was held in Stockholm, only 11 countries had environmental machineries. Two decades later, nearly all countries of the world had similar institutions.

For a variety of reasons, including efficient management, smaller institutions have generally been preferred, compared to humongous ones.

During the past century, a progressively reductionism approach has been applied to both knowledge and institutions. In one sense, integrated water resources management can be viewed as a nostalgic approach to a broader and more holistic way to manage water, as may have been possible in the past, perhaps half a century ago. However, since the world has moved on, water management needs to move with it.

In a fundamental sense, integrated water resources management, irrespective of the general impression prevalent in the water profession, is not holistic. This is not surprising, since most water professionals consider, explicitly or implicitly, water to be a very important, if not the most important, resource for human and ecosystems survival. Other issues such as energy, agriculture, industry or the environment do not generally receive appropriate emphasis or consideration from the water profession compared to water, although some of these issues may receive comparatively more attention than the others.

Increasing Complexities of Natural Resource Management

If integrated water resources management is considered essential by the water profession, other disciplines can justifiably promote very similar concepts such as integrated energy management, or integrated agricultural management, or integrated environmental management or integrated rural development. Such terminologies already exist at present, even though promotion of integration in these areas has received significantly less attention or emphasis compared to water. Unfortunately, in a complex and increasingly interdependent world, issues such as water, energy, agriculture, the environment or rural development are becoming increasingly interrelated and interdependent. Accordingly, integrated management of any one of these resources is not technically possible and institutionally and managerial feasible, because of accelerating overlaps and interlinkages with the other resource and development sectors. Developments in the water area invariably affect management of resources such as energy, agriculture or ecosystems, and the developments in these resource areas, in turn, affect water, both directly and indirectly.

As an example, let us consider the issue of water and energy interrelationships. The water profession has mostly ignored energy, even though in many ways water and energy are closely interlinked. For example, water not only produces energy (hydropower), but also the water sector is a prodigious user of energy. Accordingly, in a country such as India, hydropower accounts for slightly over 20% of electricity generated, but the water sector in turn 'consumes' a similar amount of India's electricity. In Mexico, the water sector uses an even larger percentage of national electricity generation. Furthermore, no large-scale electricity production, be it thermal, nuclear or hydro, is possible without water. In some countries such as France, the biggest user of water is not agriculture, but the energy industry. Thus, it simply is not possible to consider water resources management in an integrative manner without reference to energy, or integrated energy resources management without considering water. In other words, conceptually, technically and managerially, it is not possible to consider parallel efforts which will focus exclusively on integrated management of water or energy as a single resource, because of their inherently extensive and intensive overlaps and interlinkages.

Since water and energy are closely interrelated, integrated water resources management per se would contribute to 'unintegrated' energy management, since these two resources have many common factors in terms of planning, operation and management, which are sometimes mutually exclusive. Both of these two resources cannot be separately planned in an 'integrative' manner, irrespective of how integration is defined. Optimizing the benefits of integrated water resources management, even if this can be operationally achieved by a miracle, will not result in the maximization of the benefits of integrated energy management or vice versa. There will be substantial trade-offs, both positive and negative, for any such management approaches for these two resources in an independently integrated manner.

It can be conceivably argued that if water and energy cannot be managed in an integrative manner independently, perhaps these two resources can be managed together as integrated water and energy resources management. This is also not a practical solution because while there are significant interlinkages between water and energy, the processes available at present for their overall management are very different, and the expertise required to manage these two resources efficiently is also very different. Furthermore, institutionally, if these two resources are combined under one umbrella, for most countries it will result in a large and unmanageable institution, which is likely to be both undesirable and counterproductive. In a few countries, at least institutionally, water and energy are managed by the same governmental ministry. These countries are comparatively small, and thus the management of these two resources by one single institution may still be feasible. However, this is not possible for large- to medium-sized countries such as Brazil, China, India, Mexico, Nigeria or South Africa.

If the current global institutional arrangements for the management of water and energy resources are analyzed, they are often somewhat arbitrary. For example, hydropower in some countries such as Brazil, India, Mexico or Turkey is placed within the mandate of a separate ministry, and/or institutions, which means that the Ministry of Water has somewhat limited responsibility as to how hydropower projects are planned, operated and managed. In some other countries, the Water Ministry is responsible for hydropower, even though hydropower contributes to a very significant percentage of national electricity generation. Thus, there is no simple, elegant and universal solution in terms of integration, a fact that has been consistently ignored by the proponents of integrated water resources management. It is also interesting to note that in a country such as Canada, the word 'hydro' is synonymous with electricity, even though water and electricity are managed very differently, both technically and institutionally, at national and provincial levels.

Irrespective of whether hydropower is located institutionally within the Ministry of Energy or Water, it ensures that neither water nor energy can be managed on an integrated basis. Integration requirements, if all these can be achieved, for each of these resources are likely to be different. What is thus needed is not integration in terms of management of these two resources, but close collaboration, cooperation and coordination between the two institutions, as well as other public and private sector institutions associated with their development and management. In a real world, such collaborations are unfortunately limited, and often somewhat ad hoc. They also vary with time, even for the same country. One is reminded of Voltaire's assertion that "best is the enemy of good". The 'best' approaches for integrated water management and integrated energy management may not be compatible. What we can strive for is a 'good' solution which could result in acceptable management practices for both water and energy in a coordinated manner.

The problem becomes even more complex since it is not only the energy sector that is closely linked to water, but also other economically important sectors such as agriculture, the environment, industry or tourism. Globally, the agricultural sector is the largest user of water. Therefore, neither agriculture nor water can be managed in an 'integrated' way without considering the other. The issue becomes even more unmanageable if parallel efforts are made to manage water, energy, agriculture, industry, and/or environmental sectors in an integrated manner however the word integrated is defined. Thus, integrated water resources management, from an initial and somewhat superficial view, may appear to be a holistic approach, but on deeper consideration, it still ends up as a reductionist approach, but perhaps at a somewhat higher level.

Accordingly, integrated management of a specific resource such as water cannot simply be considered to be a holistic approach. It can be argued that it may be possible to manage two or more natural resources by combining their management processes through one common institution. Past experiences indicate that this is generally neither a practical nor efficient solution. A good example is what happened in Egypt during the 1970s, when the two separate Ministries of Irrigation and Agriculture were combined to form one single institution. The expectation was that this combined entity would manage these two sectors more rationally and efficiently than what had happened in the past. The Minister of Irrigation, who was probably one of the most dynamic and competent Ministers of Irrigation that Egypt had ever had since President Nasser's Revolution in 1952, became the minister of this new enlarged institution. In spite of his determined and strenuous efforts, it was simply not possible to manage the combined Ministry efficiently or integratively. After a very short period, the management experimentation was reversed: irrigation and agriculture became two separate ministries again. This practice has continued ever since, even though the names of the Irrigation Ministry of Egypt were changed twice subsequently. In spite of the name changes, this ministry has basically remained a water management institution, just as in the vast majority of the other countries of the world.

Additional Constraints to Implementation

In a real world, integrated water resources management, even in a limited sense, becomes difficult to achieve because of extensive inter- and intra-ministerial turf wars and bureaucratic infighting. In addition, the legal regimes (for example, national constitutions in countries such as Canada, India and Pakistan) make integrated management of any single resource very difficult. Integrated management of two or more resources by institutions that have been historic rivals is an almost impossible task.

It should also be noted that water has linkages to all development sectors and social issues such as poverty alleviation and regional income redistribution. It is simply unthinkable and totally impractical to bring them under one roof in the guise of integration, irrespective of how integration is defined. Such integrations are most likely to increase the complexities of managing the resources, instead of solving them.

Some have argued that integrated water resources management is a journey and not a destination, and the concept provides only a road map for the journey. However, the problem with such a simplistic reasoning is that in the area of water management, we are long on road maps, but very short on actual directions or competent drivers! Equally, road maps may be useful, but in order to use them we need a starting point and a destination. Without knowing the starting point and the destination, road maps are of very limited use since one is mostly likely to be all over the place. Another problem of using a road map analogy for integrated water resources management is that we do not even know where we wish to go, except in a very vague manner, and since we have no idea as to how to identify the final destination, we would have no idea when we have reached that destination, even if we reach the destination by some miracle. Not knowing the destination, it is not possible to decide if we are travelling in the right direction or the probability of reaching the right end. In the final analysis, it is not very helpful to be long on vague and unimplementable concepts but short on their implementation potential, as has been the case thus far for integrated water resources management.

There are also some negative implications of integrated water resources management, which, for the most part, are not being seriously considered.

Already, in a few countries, there are indications that the main national water institution is trying to take over other water-related institutions in the name of more effective integration. The implicit assumption is that such integration of water-related institutions will contribute to integrated water resources management. However, even if this was possible, it is unlikely to be an efficient and socially desirable approach since different institutions have different stakeholders and interests, and this diversity is a component of any democratic process. The consolidation of institutions, in the name of integration, is likely to produce more centralization, and reduced responsiveness of such institutions to the needs of the different stakeholders, which is not an objective that the present societies and international institutions prefer. Water management must be responsive to the needs and demands of a growing diversity of central, state and municipal institutions, user groups, the private sector, NGOs and other appropriate bodies. Concentration of authorities into one, or fewer, water institutions could increase bias, reduce transparency and proper scrutiny of their activities.

In addition, objectives such as increased stakeholders' participation, decentralization and decision making at the lowest possible level are unlikely to promote integration at a higher macro level, however the integration process is defined. Under most conditions, especially for macro- and meso-scale water policies, development objectives such as stakeholders' participation and a bottom-up approach at the micro-level are often unlikely to contribute to 'integration' at higher levels. This has been repeatedly observed in many developing countries such as India and Bangladesh. A variety of trade-offs between these development objectives will be necessary, since these objectives often are not mutually exclusive.

Integrated water resources management, like other similar concepts (e.g. integrated rural development, or integrated area development), has historically run into very serious difficulties in terms of their implementation. Conceptually these integrated concepts may be easy to understand, at least initially, but the world is complex, and many concepts, irrespective of their initial attractiveness and simplicity, cannot be applied to solve increasingly complex and interdependent issues and activities (Biswas & Tortajada, 2004). Even after more than half a century of existence, it has not been possible to find a practical framework that could be used for the integration of the various issues associated with water management. There is absolutely no evidence at present, irrespective of the widespread international rhetoric of the past 15 years, that this situation is likely to change in the foreseeable future.

Conclusions

It is argued that integrated water resources management has become a popular concept in recent years, but its track record in terms of application to more efficiently manage macro- and meso-scale water policies, programmes and projects has been dismal. Conceptual attraction by itself is not enough.

It should be noted that extensive analyses and research carried out at the Third World Centre for Water Management indicate that on a scale of 1 to 100 (1 being no integrated water resources management and 100 being full integration), one is hard pressed to find even a single macro- or meso-level water policy, programme or project anywhere in the world that can be given a score of 30, based on medium- to long-term performance. Indeed, it is a very dismal implementation record for a concept that has been around for nearly two generations.

Concepts and paradigms, if they are to have any validity and usefulness, must be implementable so that better and more efficient solutions can be obtained. Not only is this not happening at present for integrated water resources management, but also there are no visible signs that the situation is likely to change in the foreseeable future.

It is also necessary to ask a very fundamental question: why it has not been possible to properly implement a concept that has been around for some two generations in the real world for macro- and meso-level water policies, projects and programmes? Another important question that needs to be answered is that is the concept of integrated water resources management an universal solution, as its numerous proponents have consistently claimed, or is it a concept that has limited implementation potential, irrespective of its initial conceptual attractiveness and current popularity? Unless the concept on integrated water resources management can actually be applied in the real world to demonstrably improve the existing water management practices, its current popularity and extensive endorsements by donor institutions will unquestionably be a limited-term phenomenon, which will become irrelevant on a medium- to long-term basis.

In addition, the world is heterogeneous, with different cultures, social norms, physical attributes, skewed availability of renewable and non-renewable resources, investment funds, management capacities and institutional arrangements. The systems of governance, legal frameworks, decision-making processes and types and effectiveness of institutions mostly differ from one country to another, and often in very significant ways. Accordingly, and under such diverse conditions, one fundamental question that needs to be asked is that is it possible that a single paradigm of integrated water resources management can encompass all countries, or even regions, with diverse physical, economic, social, cultural and legal conditions? Can a single paradigm of integrated water resources management be equally valid for an economic giant like the United States, technological powerhouse like Japan, and for countries with very diverse conditions such as Brazil, Bhutan or Burkino Fasso? Can a single concept be equally applicable for Asian values, African traditions, Japanese culture, Western civilization, Islamic customs and the emerging economies of Eastern Europe? Can any general water management paradigm be equally valid for monsoon and non-monsoon countries, deserts and very wet regions, and countries in tropical, sub-tropical and temperate regions, with very different climates, institutional, legal and environmental regimes? The answer is most probably to be an emphatic 'no'.

What is now needed is an objective, impartial and undogmatic assessment of the applicability of integrated water resources management. Unfortunately, most of its current promoters have a priori assumed that this concept will automatically make the water management processes and practices ideal. Equally, the proponents of this concept have already spent so much time, energy and resources that they are mostly very reluctant to consider, let alone admit, at least in public, that the emperor may not have any clothes.

What is most likely happen in the coming years is that both the donors and the developing countries will finally appreciate the non-implementability of this concept. Based on past experience, its promoters are unlikely to admit that the concept has not worked in the past, is not working at present, and is highly unlikely to work in the future for a rapidly changing world. Accordingly, the most likely scenario of the future will be that its past and present promoters will gradually start downplaying the strong rhetoric of integrated water resources management, and start focusing on the 'ends' of water management rather than exclusive emphasis on only one of its 'means', as has been the case in recent years. A careful analysis indicates that a few international and national institutions, which have actively promoted this concept earlier, have already started to downplay it. This trend is likely to accelerate in the future.

The current evidence indicates that irrespective of the current popularity of the concept, its impact to improve water management has been, at best, marginal. It may work for micro-scale projects, but there is absolutely no evidence from anywhere in the world that it will work for macro- or meso-scale policies, programmes and projects on a long-term basis. A cynic might even say that many in the water profession mostly sit in watertight compartments, but preach integrated approaches to water management. Perhaps, the salutary caution of Harold Macmillan, the former Prime Minister of the UK, is appropriate in the current context: "After a long life I have come to the conclusion that when all the establishment is united, it is always wrong!" Is it possible that integrated water resources management falls squarely within this cautionary statement of this remarkable statesman?

References

Asian Development Bank (2007) *Asian Water Development Outlook 2007* (Manila: Asian Development Bank).

Biswas, A. K. (Ed.) (1978) *United Nations Water Conference: Summary and Main Documents* (Oxford: Pergamon Press).

Biswas, A. K. (2001) Water policies in the developing world, *International Journal of Water Resources Development*, 17(4), pp. 489–499.

Biswas, A. K. (2006) *Challenging Prevailing Wisdoms: 2006 Stockholm Water Prize Laureate Lecture* (Stockholm: Stockholm International Water Institute). Available at www.thirdworldcentre.org

Biswas, A. K. & Tortajada, C. (2004) *Appraising the Concept of Sustainable Development: Water Management and Related Environmental Challenges* (Oxford: Oxford University Press).

Biswas, A. K., Varis, O. & Tortajada, C. (2004) *Integrated Water Resources Management in South and Southeast Asia* (Oxford: Oxford University Press).

Biswas, A. K., Braga, B. P. F., Tortajada, C. & Palermo, M. (2008) Integrated water resources management in Latin America, *International Journal of Water Resources Development*, 24(1), special issue.

Global Water Partnership (2000) *Integrated Water Resources Management*. TAC Background Papers No. 4, p. 22 (Stockholm: GWP Secretariat).

Global Water Partnership (2003) *Integrated Water Resources Management Toolbox, Version 2*, p. 2 (Stockholm: GWP Secretariat).

Hall, S. S. (2003) *Merchants of Immortality* (Boston: Houghton Mifflin).

McDonnell, R. (2008) Challenges for integrated water resources management: how do we provide the knowledge to support truly integrated thinking?, *International Journal of Water Resources Development*, 24(1), pp. 131–143.

Integrated Water Resources Management: A 'Small' Step for Conceptualists, a Giant Step for Practitioners

LUIS E. GARCÍA

Introduction

In the Americas as in many other parts of the world, integrated water resources management (IWRM) originated in the need to find a reasonable compromise among the various competing uses of water when quantity and/or quality conflicts arose, because demand was larger than supply. Thus, it was not initially of much concern to the water-rich countries of the region. However, in the 1990s planners in some countries of Latin America and the Caribbean embraced the movement towards a more integrated view of making government, considering the mutual effect of the different economic, social and environmental sectors. Water was included in this and integration moved out of the water resources realm into the realm of other sectors and actors, outside the traditional water resources community. Many financing and technical assistance organizations and, most of all, the success of the Second World Water Forum of The Hague in 2000—making water the business of all and not only the business of water resource managers—reinforced this internationally. These days, it can be said that practically no national, regional or international organization in the Americas from Mexico to Chile, including Canada, fails to promote 'some' IWRM concept.

However, according to the Director of the Ground Water Institute in Pune, India, "IWRM is easy to talk about, but hard to implement". This acquires special meaning when 'the' IWRM concept is introduced in legislation in some countries and thus, it is required by law, with penalties for non-compliance. The reason is that the concept, despite the efforts of many to clarify the issue, represents many things to many people and accepts many definitions. In the extreme, some interpretations of the concept relate more to the exercise of government than to water and are more concerned with the integration of the water resource with other resources rather than on managing water itself, and also incorporates technical, institutional, social, environmental and political aspects. Thus, it is interrelated to practically all human activities.

In the 2002 Johannesburg Conference, the countries pledged to approve IWRM and efficiency plans by 2005. How are the Americas coping with this challenge? By 2005, according to a GWP report, 95 countries worldwide (16 from Latin America) showed some progress in this regard (in Latin America one showed good progress, 10 had taken some steps, and five were in the initial stages). Since then, two (Costa Rica and Mexico) have almost finished their IWRM strategy, and one (Dominican Republic) is in the initial stages. This paper discusses lessons to practitioners learned from the Mexico, Costa Rica and Guatemala cases, from a practical point of view.

In the Beginning . . .

When searching for the origins of the IWRM concept, it is not uncommon to find in the literature cites placing it at the Dublin Conference in 1992, at the Mar del Plata United Nations Conference on Water in 1977, or even after the creation of the Global Water Partnership (GWP) in 1996. However, some may state that it has been around for much longer than that, and the author tends to agree with this perception.

It was first encountered by the author in the mid-1960s, although it surely must have been around for some time before that, during a discussion on water use conflicts at the University of California in Berkeley, in the form of the 'Reasonable Use Principle' (Todd, 1965). When more than one use or user needed the same water supply source, needed the same water volume stored or released from a reservoir at different times, or changed the water quality characteristics of a given receiving water body, conflicts appeared. Every use or user wanted to maximize the water use benefits simultaneously, but of course, that was not possible. It must have been realized then, that maximizing the benefits for one use or user interfered with others, and the only way to solve or minimize those conflicts would be to reach a compromise among users. If placed within a context of a watershed, river basin or region, the 'best' compromise would be that which maximized the benefits for the whole (the watershed, river basin or region) and not for any of its parts (the individual uses or users). Interestingly, the same principle was heard again during a lecture in Buenos Aires (Sadoff, 2003), which defined IWRM as maximizing the benefits for the system and not any of its components. As illustrated by the hypothetical example in Figure 1, a combination of hydropower generation upstream and irrigation downstream would be better for the whole system than maximizing the benefits for upstream irrigation only.

Certainly, the concept of IWRM described above has been present in the minds of Latin American water resources managers during the last 40 years. In general, the region moved from the project-by-project approach prevalent in the 1960s to a sub-sectoral approach in the 1970s and mid-1980s, when irrigation, water supply and energy master plans were in vogue (IDB, 1998). But apart from some attempts in countries such as Colombia to capitalize the successful TVA approach, not many examples of real world applications of IWRM can be easily found.

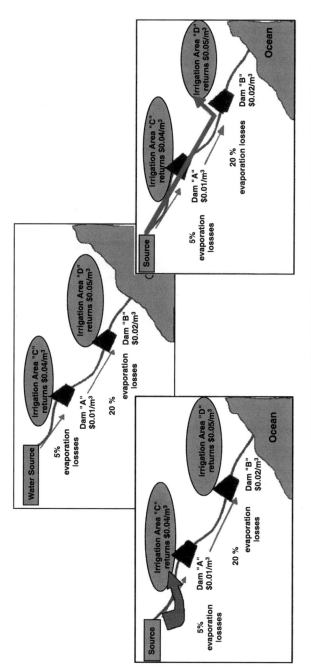

	VALUE FOR USES UPSTREAM	VALUE FOR USES DOWNSTREAM	VALUE FOR SYSTEM	
Option 1 Irrigation upstream only	$0.04/m³	Minimal	$0.038/m³	Upstream use better
Option 2 Upstream hydropower + downstream irrigation	$0.01/m³ Lower	$0.05/m³ Much higher	$0.048/m³ Higher	Higher system value with benefits to both uses

Figure 1. Reasonable use principle. *Source:* C. Sadoff (2003) adapted by the author.

The reason may be that water quantity and/or quality conflicts were not the norm in the region during much of the last half century, because the Americas is the water-richest region of the world, with more than 55% of its total renewable freshwater (García & Aguilar, 2006). With a mean annual precipitation of 1084 mm, the total renewable water resource in the region amounts to nearly 24 000 km^3 per year, but exhibiting great heterogeneity. It ranges from approximately 90 km^3 per year in the Caribbean, to around 700 km^3 per year in Central America, to over 6000 km^3 per year in North America, to slightly over 17 000 km^3 per year in South America. It seems there is water for everyone as the following average figures of m^3/person/year show: approximately 2200 in the Caribbean, 15 000 in North America, 18 000 in Central America, and 47 000 in South America, although individual countries show great variation. However, while the global figure of withdrawals reaches an average of 8.7% of the total available resources in the Americas, it is as high as 15% in the United States, Mexico and some Caribbean countries, but as low as less than 1% in several countries of Central and South America, resulting in an average of only 3.2%.

As a result, the need to find a reasonable compromise among the various competing uses of water was not a pressing one in the region during the past half century, and the application of the 'Reasonable Use Principle' version of the IWRM concept was not of much concern to the water resources planners in many of the water-rich countries of the region, except in those areas such as Northwestern Mexico, Northern Chile, parts of Bolivia and Peru, or Northeastern Brazil, where water demands gradually became greater than the supply.

However ...

The generalization of awareness about climatic variability, coupled with increases in population and urbanization, and renewed concerns for the environment started to change this picture during the last decade. The Americas is a region with great variety of climate conditions. It has ice caps and glaciers, snowy mountain peaks, four-season weather in some areas and two-season dry and wet cycles in others. It has humid rainforests as well as arid and semi-arid areas. Some areas receive more than 6000 mm of rainfall in a year, whereas others, such as the Atacama Desert, are the driest in the world.

The population is growing and distributing itself not precisely in the water richest areas. The population annual growth rate in the Americas is 1.22%, similar to that globally. Although it is lower in some areas, such as the Caribbean (1.03%) and North America (1.05%), it is higher in South America (1.33%) and Central America (2.21%). In 2004 the population of Latin America and the Caribbean reached 545 million, representing 63% of the population of the Americas, and 8.6% of the world. About 48% of the world population lived in cities in 2004. However, the urbanization reached 52% in Central America, 60% in the Caribbean, 75% in Mexico, and 81% in South America.

A 'Small' Conceptual Step ...

New concerns about global warming and the environment in general and a renewed interest for sustainable development after the so-called 'lost decade' of the 1980s, fuelled a movement in the 1990s towards a more integrated view of making government, considering the mutual effects of the different economic, social and environmental variables. The success of the Second World Water Forum of The Hague in 2000 in making water the business of all, and not only the business of water resources managers—supported by many financing and technical assistance organizations—reinforced this internationally. Integration moved out of the water resources realm into the sphere of other sectors and actors outside the traditional water resources community. Today, it can be said that almost no national, regional or international organization in Latin America and the Caribbean, from Mexico to Chile, fails to promote 'some' IWRM concept. This is by no means a small step conceptually, but it is a 'small' step if it compared to what would be needed to bring it to practice.

Giant Steps Needed for Practitioners ...

Perhaps the first giant step is to find who the practitioners are. The second step would be for practitioners to come to a universal understanding as to what IWRM means to them in practice. The third and subsequent steps would be for them to actually use IWRM to improve water services for the end users. The first two would be necessary, but not sufficient, conditions for IWRM implementation, unless the third one is met. At present, the most commonly looked at definition of IWRM is that of the GWP (2000):

> IWRM is a process which promotes the co-ordinated development and management of water, land and related resources, in order to maximize the resultant economic and social welfare in an equitable manner without compromising the sustainability of vital ecosystems.

However, this is a conceptual definition and despite the efforts of the GWP and many others to clarify the issue, IWRM conceptually represents many things to many people and accepts many definitions. More than 35 definitions have been found in the literature (Biswas, 2004) in support of integration of the natural system and the human system, water and land resources, surface and groundwater, river basins and coastal zones, water volume and water quality, and upstream and downstream uses and users. The need for integration throughout the many sectors of the economy has also been pointed out, as well as the need to consider the macro-economic effects of water projects and the participation of all stakeholders. On the one hand, some interpretations of the concept relate more to the exercise of government than to water and are more concerned with the integration of the water resource with other resources rather than on managing water itself. On the other hand, IWRM has been defined essentially as environmental management (Hofstede, 2006) and considered as a complement to, complemented by, or part of the ecosystem approach which in itself is a holistic approach. Others incorporate technical, institutional, social, environmental and political aspects, making it interrelated to practically all human activities.

So, who are the practitioners and will they ever come to a universal understanding as to what IWRM means in practice? The following examples from Latin America may help to answer this question.

Examples from Latin America and the Caribbean

The process of preparation of the Regional Report from the Americas to the Fourth World Water Forum held in Mexico in March 2006 helped to review the many challenges facing the region regarding water resources management. It also helped to highlight many accomplishments for which the region can feel proud, and also evidenced several issues on which there is no consensus and maybe even one or two controversies (García & Aguilar, 2006). The report reflected different points of view about issues such as the importance of water demand management versus water resources development; water as an economic good versus water as a human right; the effect of international trade agreements on national water rights; the role of infrastructure, especially dams; private sector participation in the provision of basic water services; the real meaning of payment for environmental services; and last but not least, the implementation of IWRM with debates ranging from the meaning of IWRM itself to the usefulness or practicality of its implementation.

The Inter-American Development Bank IWRM Strategy

The Inter-American Development Bank (IDB) is the oldest of the Regional Banks and the major development lender in the region. Since the Bank was founded more than 40 years ago, it has provided financing (an average of more than US$900 million per year totalling close to US$40 billion in the period; approximately 18% of its lending portfolio) to water-related projects and activities. Since then, this category has remained a common denominator in the Bank's financing portfolio. The social role as well as the productive role of water in economic growth and poverty alleviation has received due importance, as the Bank has financed many irrigation, drainage, hydropower and other water-related projects, as well as flood control projects.

During the first 25 years of the Bank the emphasis in the region was on infrastructure on a sub-sectoral project-by-project basis, with few multi-purpose developments. Investments in water supply and sanitation dominated during the Bank's early years and hydropower investments started to increase gradually in the 1970s, becoming dominant in the 1980s, and ultimately declining from then on. Investments in irrigation and drainage reached a peak during the second half of the 1970s and also declined during the second half of the 1980s. In the 1990s, the Eighth General Increase in the Resources of the IDB, known as the 'Eight Replenishment', or IDB-8, called for more attention to be paid to the socio-economic and environmental sectors. This was reflected mostly in water resources by new trends in financing of water projects with an increased concern for the watersheds and people living in them, for the quality of receiving waters, for management as opposed to development of the water resource, and for integrated water resources planning (IDB, 1994).

Following the mandates of IDB-8, the Bank recognized that meeting local and provincial-level objectives, together with national objectives, were important for development. It was also recognized that infrastructure alone could not solve the problems, and that sometimes these problems were aggravated by neglecting other equally important social and environmental variables. Therefore, in 1990 the Bank strengthened a process to incorporate these variables in the financing of water related projects, and in 1998 the Board of Directors approved a strategy for IWRM to foster the application of the mandates of IDB-8 to the water sector by focusing on mainstreaming the principles of IWRM in its water-related projects.

Fundamentally, the IDB IWRM strategy focused on actions directed towards three levels as depicted in Figure 2. At the constitutional level, actions to help the countries in drafting, approving and applying basic water legislation and IWRM policies and strategies were promoted to help create an enabling environment. At the associative level, the river basin or watershed approach was fostered and attention was focused on integrated

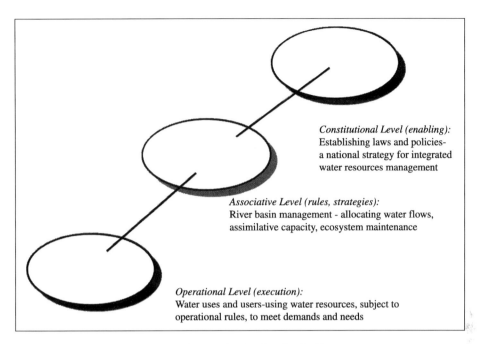

Constitutional Level (enabling):
Establishing laws and policies-
a national strategy for integrated
water resources management

Associative Level (rules, strategies):
River basin management - allocating water flows,
assimilative capacity, ecosystem maintenance

Operational Level (execution):
Water uses and users-using water resources, subject to
operational rules, to meet demands and needs

Figure 2. Levels of action for IWRM

watershed management, including river basin organizations. At the operational level, financing of water supply and sanitation, irrigation and drainage, and hydropower projects were continued but with a new emphasis on sub-sector modernization, especially in the water supply and sanitation sub-sector.

An institutional model for IWRM was proposed that was not very different to the one also being proposed by the World Bank (Figure 3). This model involved the separation of four distinct functions using regulatory bodies: the water allocation among competing uses function; the provision of water as a social service and as a production factor function; the conservation and/or improvement of the quality of the environment function; and the supply of reliable and timely provision of basic water resources and other related information function.

A recent review of the IDB 1990–2005 water related portfolio (García & Ortiz, 2006, p. 46) pointed out that "support for water management is especially evident in watershed management and planning". Efforts in this sub-sector gave due consideration to the river

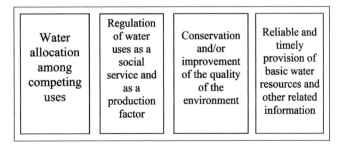

Water allocation among competing uses	Regulation of water uses as a social service and as a production factor	Conservation and/or improvement of the quality of the environment	Reliable and timely provision of basic water resources and other related information

Figure 3. Institutional model for IWRM

basin and also made contributions to governance and to the integration of the socio-cultural and environmental variables into water projects. In particular, the Bank was active in promoting an enabling environment, including support for drafting new legislation, policies, strategies and action plans for integrated water resources management in the region, mainly through technical cooperation operations and the non-financial portfolio. Despite these efforts, however, no clear evidence was found of linkages between the integrated water resources management approach and the projects in the rest of the investment portfolio. During the period under consideration, the Bank also placed special emphasis on the institutional reform and modernization of the water resources sector and on implementing IWRM in the region. However, these efforts are not yet consolidated operationally and the efforts to mainstream this approach should be strengthened.

Major Practical Hurdles

The major hurdles encountered in this effort relate to the emphasis that was placed on applying the IWRM concept at the constitutional and associative levels compared to the operational level, although the countries in general showed more interest in water services than in water resources management. In addition, initiatives and proposals for new water legislation were anything but scarce, but very few were actually approved. With the exception of cases related to the provision of water function, the achievement of institutional reforms proved to be extremely difficult in practice, especially those related to the water allocation among competing uses function. A major effort was directed towards exploring a basin-wide integrated water quality approach (Russell *et al.*, 2001). The main obstacles encountered were the need for the collection of large amounts of basin-wide reliable data (which was serious but achievable in time), the availability of reliable predictive water quality models (which existed), and the investments needed to obtain benefits from water quality improvements basin-wide (which proved to be prohibitive).

The IWRM Country Strategies

At the Summit for Sustainable Development held in Johannesburg in 2002, the countries pledged to approve IWRM and efficiency plans by 2005. The latest survey made by GWP (2006) among 95 countries showed that 21% of the countries surveyed had plans or strategies in place and 53% had initiated its formulation process. Therefore, almost 75% of the countries surveyed have met the target set out by the Johannesburg initiative.[1] According to the GWP report, in Latin America and the Caribbean:

- One country (Brazil) has IWRM plans or strategies in place.
- Ten countries (Belize, Costa Rica, El Salvador, Nicaragua, Panama, Argentina, Chile, Colombia, Peru and Uruguay) are in the process of preparing them.
- Five countries (Guatemala, Honduras, Bolivia, Paraguay and Venezuela) have taken initial steps.

Since then, two (Costa Rica and Mexico) have almost finished their IWRM strategy, and one (Dominican Republic) is in the initial stages.

Brazil has made a good effort to translate the conceptual IWRM definition posed by GWP into practical terms. In this sense, IWRM is applied in the planning process through

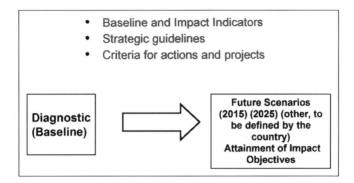

Figure 4. General steps for a strategy

consideration and evaluation of the following aspects:

- In any given river basin, multiple uses of water are considered and evaluated.
- Economic, social and environmental objectives are considered and weighed up.
- Due consideration is given to other government areas, levels, and instruments, such as national and regional plans.

This does not mean that all of them are included in the final plan, but they are considered, weighed up and evaluated in its initial stages and then only what is relevant for the case in question is included. Stakeholders are included in the decision-making process through a transparent approach.

An effort to speed up the process was initiated in 2006 by the UNEP Collaborating Centre on Water and Environment (UCC-Water) and the DHI-Water and Environment with the collaboration of IUCN. Workshops with that purpose in mind have been held in Central America, the Caribbean, the South American Andean region, and the South American Southern Cone. Although there were some hurdles, at least there is consensus with regard to the general steps needed to formulate an IWRM Strategy, as shown in Figure 4.

Major Practical Hurdles

Although these results are encouraging, some difficulties have been encountered from the operational level end-of-the-line user point of view. As with the IDB IWRM Strategy previously discussed, most of these efforts have been focused at the constitutional level and even if they are strong in concepts, few actions meaningful for the end-user have resulted so far. There is a tendency to be heavy on diagnostics but weak in solutions and a gap still exists between papers and actions. Some propose solutions to water problems that do not fit with the socio-economic and socio-politic systems prevailing in the countries for which they are proposed and, in some cases, the discussion of issues transcend the water arena into the socio-political and ideological arenas. In addition, in cases where the proposing agency is also a water user, other agencies perceive a bias, provoking strong resistance from them. In spite of the strong interest showed by the countries for IWRM in some regions, the results of the UCC-Water/DHI-Water and Environment have shown a wide variety of interpretations as to what IWRM is and how it can be applied. In the Central American and Andean Region at least, there was a strong bias towards an environmental interpretation of the IWRM concept, which seems to be taking hold across the region. There was also strong interest for the policy, strategy and legislation aspects, but a lack of clarity about how to follow those with action plans was perceived, at least among some of the participants.

IWRM Strategy in Mexico

The IWRM concept is fully accepted and has been adopted to the extent that IWRM, as defined by GWP, appears at least 37 times in the 2004 National Water Law (LAN). The Law also includes sanctions for non-compliance and much discussion has been had in the National Water Commission (CONAGUA) and other institutions regarding how to comply with such a non-operational definition. Does compliance mean everything in the definition, full integration or partial integration? Where is the interface between multisectoral water resources management, watershed management, natural resource management, environmental management and socio-political management, or is there no interface and should the holistic approach be taken? Is CONAGUA capable of that? Will it not interfere with the functions delegated by law to other institutions? These are but only a few of the many questions that prompted the National Water Commission to develop and adopt a conceptual IWRM paper and it is in the process of developing specific regional action plans.

As in Brazil, Mexico has also adopted a pragmatic translation of the IWRM concept to be applied in specific hydrologic units (such as watersheds and river basins). For water resources management to be integrated, the following processes should be included, as long as they are relevant for solving the problem under analysis:

- soil-water and water-vegetation relationships;
- economic, social and environmental aspects;
- actions from all levels of government and stakeholders;
- surface and groundwater;
- water volume and water quality;
- supply and demand oriented actions; and
- uses and/or users that compete for the same water body or source.

The 'Reasonable Use Principle' was also considered when there are various beneficial uses and users that compete for the same source or water body, stating that the solution is to maximize the benefits for the whole system (region, basin, aquifer) and not for a specific use, giving due consideration to concertation and conflict resolution techniques. However, much discussion was generated when considering which of the three sustainable development objectives should be maximized, finally agreeing within CONAGUA that economic efficiency should be sought, including social and environmental considerations as stipulations in an optimization set of equations. Such an approach has been proposed to re-distribute and increase the water use efficiency in the state of Aguascalientes, whose water supply comes mainly from one of the five aquifers existing within it. It was found that the safe yield was 300 million m^3 per year, while extractions added up to almost 500 million m^3 per year. 70% of the extractions were used for agriculture and livestock, which contributed 5% to the state GDP, while 5% were used for industry, which contributed 30% to the state GDP and 25% were used for public and urban uses, whose contribution to the state GDP was estimated to be 65%.

IWRM Strategy in Costa Rica

With support from the IDB-Netherlands Water Partnership (INWAP), the government of Costa Rica embarked on an ambitious programme to develop its IWRM master Plan in two stages. The first stage, which has been successfully completed, consisted of developing an IWRM Strategy through a highly participatory approach, to be followed by a second stage already under way to develop the Master Plan itself. This is a two-step approach that has been favoured by INWAP to bridge the paper-action gap so many times observed in similar efforts, where the strategy is not followed by actual user-oriented actions capable of being financed by the country under its budgetary and management capabilities.

It is known that Costa Rica's national policies are highly environmentally oriented and, as in many other countries in Central America, the Ministry of Environment (MINAE) has control of the water. This posed a difficulty by introducing an environmental bias that was strongly contested by other institutions such as the Costa Rican Electricity Institute (ICE), the Water Supply and Sewerage Institute (AyA) and the National Irrigation and Drainage Service (SENARA), all of which include in their legal responsibilities dealing with the water resource. Much of the discussion revolved around who would be the National Water Authority, although current Costa Rican legislation assigns that role to MINAE. However, the baseline was who would charge for water extractions and how much.

After much discussion, an agreement was reached about the distribution of responsibilities concerning the water resources functions, as depicted in Figure 3. MINAE was left in charge of the water allocation among competing uses function, the conservation and/or improvement of the quality of the environment function, and part of the provision of reliable and timely provision of water resources and other related information. The other Ministries and service providers such as ICE, AyA and SENARA were left in charge of their respective social and productive water uses, regulated by a single regulatory body, and part of the provision of information function.

During the development of the IWRM Strategy, a new Water Law was also being discussed in Congress and that posed simultaneously a problem and an advantage. The problem was that both the proposed strategy and law included policy guidelines that, if not concerted, could point the country in different directions. The advantage was that the simultaneous discussion of both instruments provided an opportunity for concerted effort and coordination, which was fully taken advantage of by both teams. The strategy was finally approved by the government but, unfortunately, the approval of the new Water Law is still on hold.

Another issue that had to be solved during the process was that market-based approaches were not politically nor socially favoured in Costa Rica at the time. The objective of economic efficiency with social and environmental considerations as constraints in the optimization was not fully appreciated, and requiring careful wording pointing towards that direction, but maintaining the favoured social and environmental prevalence.

IWRM Strategy in Guatemala

This effort was started after Costa Rica, also supported by INWAP, and benefited from the Costa Rican experience. The same two-step approach was used and at present the IWRM Strategy is being developed. However, there are important differences between the two countries, both in approaches and in what is considered important. In contrast to Costa

Rica, Guatemala has a high indigenous population whose conceptualization of the meaning of water and water rights is different.

First, the project was favoured by the Planning and Programming Secretariat of the Presidence (SEGEPLAN), who was interested in rationalizing the public expenditures in water projects and as such, was void of any bias towards any water user. Unfortunately, the project had to compete with an existing effort by the Ministry of Agriculture, supported by IDB, to develop a water resources strategy and with an effort by the Ministry of Environment to coordinate the development of a government water policy through the National Water Commission (CONAGUA). In contrast to Mexico, where CONAGUA is a de-concentrated institution with a rank and organization close to a Ministry, in Guatemala CONAGUA is a Committee formed by representatives from the many institutions that deal with water.

Until the creation of the Ministry of Environment, the Ministry of Agriculture had traditionally been the water authority in the country and, although its effort towards an IWRM strategy was discussed in CONAGUA, the feeling of bias towards water use in agriculture and livestock favoured SEGEPLAN's effort. In addition, the water policy being coordinated by the Ministry of Environment considered only the government sector, and SEGEPLAN wanted to also include the private sector and civil society.

As a result, when it began, the SEGEPLAN project focused on how water could help achieve the objectives of four national plans started by the government with the aim of improving living conditions in the poorest municipalities (*Guate Solidaria* in Spanish), foster economic growth (*Guate crece*), improve competitiveness (*Guate compite*), and protect the environment (*Guate verde*). However, compared to Costa Rica, the effort has so far been concentrated mainly within SEGEPLAN and the project team. Nevertheless, as a result of guidelines given during a presentation to the Cabinet, consultations with private sector stakeholders and civil society representatives, including indigenous peoples, are being planned.

IWRM Local Actions in the Fourth World Water Forum

The Fourth World Water Forum held in Mexico in March 2006 had five thematic areas: water for growth and development; implementing IWRM; water supply and sanitation for all; water management for food and the environment; and risk management. The activities consisted mainly of presentations of local actions regarding those five areas.

In its final report (WWC-CONAGUA, 2006), the Fourth World Water Forum Secretariat states that of the 1477 local actions submitted to the Forum from all over the world, the greatest number (31%) were on the theme 'Implementing IWRM'. Second (27%) were local actions associated with the theme 'Water Supply and Sanitation for All'. Local actions related to the theme 'Water for Growth and Development' were third (18%), those related to 'Water Management for Food and the Environment' were fourth (17%), and those related to 'Risk Management' were last (7%).

The Americas region submitted 885 local actions (García & Aguilar, 2006) and the thematic ranking was the same as the global ranking. 'Implementing IWRM' and 'Water Supply and Sanitation for All' were also in first and second place (28% and 24%, respectively). 'Water for Growth and Development', 'Water Management for Food and the Environment', and 'Risk Management' followed, with 22%, 21%, and 5% respectively.

Of course, this can be interpreted as revealing the importance of IWRM as the foundation for its proper utilization, both globally as stated in the Forum's final report, and in the Americas region. However, there is a strong temptation to ask the following biased questions: if globally, 31% of local actions were about IWRM, does that mean that 69% were not? Similarly, in the Americas, if 28% of local actions were about IWRM, does it mean that 72% were not? If all the local actions associated with water for growth and development, water supply and sanitation, water management for food and the environment, and risk management did not need to use the IWRM approach, for what purpose was then IWRM used?

The Operative Committee of the Americas (OCA) that coordinated the participation of the region in the Forum named an independent Committee to rank the local actions submitted for the purpose of being included in the Regional Report of the Americas. Of the 11 highest ranked IWRM local actions, four were associated with reforestation in upper watersheds and conservation of water resources; two were associated with bi-national arrangements for watershed management; one was associated with educational programmes for IWRM; one dealt with the management of invading plant species in water bodies for their use by local artisans; one was related to bio-treatment of waste water; one described the agreements reached in Mexico's Lerma-Chapala basin for the distribution of its water resources; and one described the Mexico National Water Plan.

Apart from the last two, all of the others seemed to describe actions that had been taking place before under a different name, and it is hard to rationalize how they could belong to a group described as the foundation for a proper utilization of water resources for growth and development, water supply and sanitation, food and the environment or risk management. However, the common denominator is that all involved the coordinated participation of many actors (local, municipal, state and federal).

Troubling Questions for Practitioners

Given the previous discussions, there is a temptation to ask the following questions:

- Is IWRM applicable mainly at the institutional and associative levels and only towards the creation of an enabling environment, and not at the operational level? (Figure 2).
- Is it now *I*WRM instead of *IWRM*, meaning that the emphasis is in integration instead of water resources management?
- Should the emphasis be on water resources management instead of integration?
- Should the importance of integration diminish as the actions move from the constitutional level to the associative and to the operational levels, or vice versa?
- Is IWRM now more related to the exercise of government and governance than to water management?
- Is IWRM more concerned with the integration of the water resource with other resources rather than within the water resource itself?
- Has the incorporation of institutional, social, environmental and political aspects migrated the discussion to issues within these other sectors, rather than within the water sector itself or its relationships with the other sectors?
- Has the discussion of IWRM moved out of water into the socio-economic, socio-political and ideological arenas?

These are indeed troubling questions for practitioners. There should be hope that the answers are not. There should be hope that the answers will help them take the third and subsequent giant steps mentioned earlier, so that IWRM is used to achieve practical results for the end-users of water.

Note

1. Other surveys have also been made, with similar results.

References

Biswas, A. K. (2004) Integrated Water Resources Management; a reassessment, *Water International*, 29(2), pp. 248–256.

García, L. E. & Aguilar, E. (2006) Regional Report of the Americas for the Fourth World Water Forum, Mexico, March.

García, L. E. & Ortiz, S. (2006) *Water Resources. Support from the Inter-American Development Bank Group, 1990–2005* (Washington DC: Inter-American Bank).

GWP (2000) Integrated Water Resources Management, TAC Background Paper No. 4 (Stockholm: Global Water Partnership, Technical Advisory Committee (TAC)).

GWP (2006) Setting the Stage for Change. Second informal survey by the GWP network giving the status of the 2005 WSSD target on national integrated water resources management and water efficiency plans, February 2006 (Stockholm: GWP).

Hofstede, R. (2006) Ecosistemas, Cuencas Hidrográficas y Gestión Integral del Agua. UNEP workshop in support of the Johannesburg Directive, Andean Region, Quito, Ecuador, 18–19 October.

IDB (1994) *Eight Increase in the Capital of the Bank (IDB-8)* (Washington DC: Inter-American Development Bank).

IDB (1998) Strategy for Integrated Water Resources Management. IDB Strategy Paper No. ENV-125, December (Washington DC: Inter-American Development Bank).

Russell, C. S., William, J. V., Clark, C. D., Rodríguez, D. J. & Darling, A. H. (2001) *Investing in Water Quality: Measuring Benefits, Costs and Risks* (Washington DC: Inter-American Development Bank).

Sadoff, C. (2003) Presentation at the IWRM Course for South America. IDE World Bank/IDB, Buenos Aires, Argentina.

Todd, D. (1965), Notes from lectures at the University of California, Berkeley.

WWC-CONAGUA (2006) *Final Report. Mexico 2006 4th World Water Forum* (Mexico: DF).

Integrated River Basin Plan in Practice: The São Francisco River Basin

B. P. F. BRAGA & J. G. LOTUFO

Introduction

Integrated water resources management (IWRM) as defined by GWP (2003, p. 1) is a:

> process which promotes the coordinated development and management of water, land and related resources in order to maximize the resultant economic and social welfare in an equitable manner without compromising the sustainability of vital ecosystems.

This process is also described in more detail in the seminal paper by Biswas (1978). Of course, the GWP definition is too general and needs to be decoded in some more practical terms in order to be implemented in real life applications. In this paper, IWRM is a process that considers multiple water uses in a river basin, their conflicts and complementarities. It also considers multiple objectives, including economic, social and environmental objectives in planning and managing river basin. It allows the coordination with other areas and levels of government (national planning and regional planning) and promotes the involvement of stakeholders in an open decision-making process.

The recent legal and institutional reform that has taken place in Brazil in the last 10 years provides the platform for the implementation of IWRM in the country.

Law 9433 issued in January 1997 provides the elements for the consideration of multiple water uses, public participation, water permits, charges and control.

The integrated river basin plan presented here considers all these elements of IWRM in its construction. The São Francisco river basin located in the north-eastern part of Brazil, covers more than $600\,000\,\mathrm{km}^2$ and has a population of 13.3 million predominantly urban dwellers. This basin has a very active river basin committee, which is responsible for approving the river basin plan. The São Francisco River Basin Committee (SFRBC) has 60 members, 20 of whom are from public sector (federal, state and municipal level), 24 are water users organizations and 16 are from the organized civil society.

The river basin plan was developed by the National Water Agency of Brazil (ANA) with the active participation the river basin committee members and other States and federal water resources agencies.

Institutional Framework

The São Francisco river (Figure 1) is $2863\,\mathrm{km}$ long with a drainage area is $636\,920\,\mathrm{km}^2$ (8% of the national territory), and encompasses 503 municipalities of seven federative units, including six states (Bahia, Minas Gerais, Pernambuco, Alagoas, Sergipe and Goiás) and the federal district. According to last national census in 2000, approximately 13.3 million predominantly urban inhabitants live in the basin. Until very recently, studies and projects conducted in the São Francisco river basin region had neither adopted the basin as a the territorial unit of planning including its coastal zone, nor an integrated management perspective. Through a GEF/UNEP/OAS funded joint initiative with ANA, an Integrated River Basin Plan was developed. This project, launched in the year 2000, highlighted the importance of exchanging experiences and information, as well as sharing information among sub-basins of the São Francisco. The project had a major public participation component. More than 12,000 professionals of different sectors related to water resources participated in the work.

A Basin Analytical Diagnosis (BAD) and correspondent Strategic Action Programme (SAP) were produced according to the Global Environmental Facility (GEF) methodology (GEF, 2000). Because of its size, the project proposed a physiographic division of the basin in four regions: upper, middle, sub-middle and lower São Francisco basin (Figure 1). The SAP was accepted by the SFRBC as a basis for the preparation of the São Francisco River Basin Water Resources Plan (SFRBP) A working group was constituted to prepare the plan, coordinated by ANA, and consisting of representatives from a federal irrigation agency, CODEVASF, representatives of all federative units in the basin, two federal hydropower companies (CEMIG and CHESF), and members of the Ministry of Environment in charge of a federal environmental restoration programme in the basin.

During the preparation of the plan, previously conducted studies were capitalized on, such as the BAD, the SAP, and the basic studies of the Brazilian National Water Resource Plan (NWRP). ANA also prepared topic studies to improve its implementation conditions. The working group activities were followed up by two Technical Chambers (Plans and Programmes, and Permit and Charging), by Regional Advisory Chambers of the SFRBC and by its Board, in systematic meetings held in Brasilia and cities in the basin. Over this participatory planning process, the main purpose of the SFRBP was to define an agenda for the basin, including management actions, programmes, projects, civil works and priority

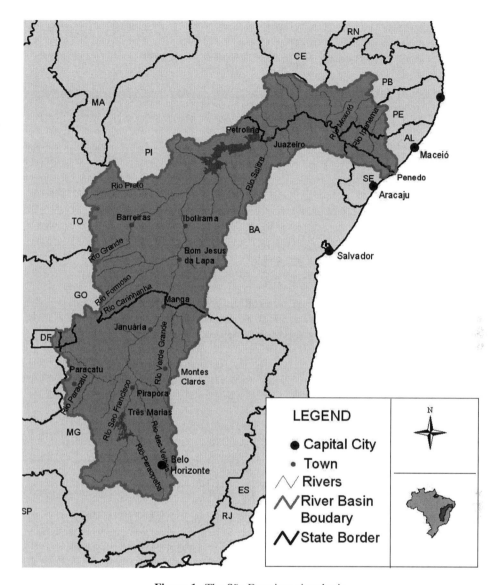

Figure 1. The São Francisco river basin

investments within a context that encompassed government agencies, the civil society, users and the many institutions that take part in water resource management.

The following specific objectives have been considered:

- to implement the Integrated Basin Water Resource Management System;
- to establish guidelines for the allocation and sustainable use of the basin water resources;
- to define a strategy for the basin hydro-environmental conservation and revitalization; and

- to propose actions and investments in water resource services and works, land use conservation and environmental sanitation.

In order to discuss the SFRBP proposal, the basin committee formed a Technical Support Group (TSG), aiming to support the Executive Secretariat and the Technical Chambers of the Committee. Essentially, the plan review conducted by the TSG got limited to two main topics: water allocation and the basin revitalization programme. These two topics were also discussed during an Evaluation Forum and in public consultations held in the four basin physiographical regions. These events were attended by representatives of the public sector, the organized civil society, users and entrepreneurs. The plan adopted the results of the public discussions entertained in the Evaluation Forum and in the two rounds of regional consultations and was approved by the SFRBC. The following outputs were envisioned:

- Module 1: Executive summary.
- Module 2: Consolidated diagnosis of the basin, and development scenarios.
- Module 3: Water allocation, classification, integrated control and charging.
- Module 4: Strategies for revitalization, recovery and hydro-environmental conservation of the basin, and investment programme.

This paper presents the main achievements of this plan, including the deliberations of the plenary session of the SFRBC especially called for the approval of the SFRBP. The deliberative characteristic of the river basin committees makes public participation more effective in Brazil.

Basin Diagnosis and Development Scenarios

The diagnosis outlined a panorama of the main characteristics of the basin, and consolidated the most relevant aspects related to the user sectors and to water availability, including a review of the potential and existing conflicts, as well as a proposal to attune the use of the water resources. This information, as well as detailed development scenarios, served as an input to the remaining phases of the Plan.

Antecedent Conditions

The plan design happened within an historical context, where multiple factors were specifically important in each phase. On the one hand, these factors represented a starting point for the organization of this work and, on the other hand, the potential and problems to be explored by the Plan. Some of these factors stand out, such as:

- the dimensions and complexity of the São Francisco river basin (Figure 1), which encompasses six states and the federal district, highlighting the importance of implementing, in the basin, a water-resource management model founded on decentralization, participation, integration and negotiation;
- issues of the political-administrative organization of the basin that show an institutional weakness, such as the many agencies that deal with development in a disorganized and sectoral manner, as well as double ownership of the water resources (seven Federation Units and the federal government), reinforce the need for integration of governmental and civil society actions to improve the legal framework and the harmony of policies;

- the existence of a slow but growing and dynamic process of social participation that has thrived since the promulgation of the 1988 Constitution, in the municipalities and sectors of the basin, especially the establishment of the SFRBC and various state committees, and sub-basins;
- the existence of persistent socio-economic paradoxes, heterogeneity, asymmetries and inequities over the time, which leads to the co-existence of regions of remarkable wealth, high demographic density, and accelerated urbanization and regions of extreme poverty, sparse population and meaningless economic growth, such as:
  - the upstream São Francisco river region, where Belo Horizonte Metropolitan Region (RMBH) is located, which accounts for less than 1% of the total basin area (638 576 km^2) and concentrates more than 3 90 0000 inhabitants, i.e. approximately 29.3% of the total population (12 796 082 inhabitants); and
  - the semi-arid region, a fragile area of water shortage and low socio-economic condition that should be subject to specific actions;
- the whole picture of the basin environmental degradation that shows a loss of biodiversity and changes in the aquatic ecosystems resulting from factors such as poor sanitation services, the construction of big dams and industrial and agricultural activities in the basin, impairing the quality of the water, indicating the need of actions of both educational and preventive nature, as far as recovery and adaptation are concerned;
- rural economic activities such as agriculture, mining activities particularly concentrated in the São Francisco river upstream region, and the basin urbanization process that promote the removal of the native vegetation and speed up the processes of erosion and silting, indicate a need for environmental recovery of the degraded areas, aiming to mitigate their impact on the water resources; and
- the presence of a major diversity of water usage, which changes the concept of multiple and concurrent uses into a fundamental approach for the management of the basin resources. Such a myriad of aspects represents the starting point for the development of the Plan, an indication of vulnerabilities and potentials to be explored in the basin, in order to harmonize the use of the water resources for water supply and dilution of the effluents, irrigation of farmable soils, generation of power, navigation, fishing, aquiculture, development of tourism activities and entertainment, and maintenance of ecosystems.

Availability and Demand for Water Resources

As far as water resource management is concerned, quantity and quality cannot be detached from each other, and this has justified the need to evaluate water availability in quantitative and qualitative terms. Q_{95} or the flow with 95% probability of being surpassed in any given year was used as the standard for surface water availability. In the São Francisco river reaches with high regulation capacity, water availability is regarded as being the regulated discharge added to Q_{95} discharge. Downstream of Três Marias reservoir, the regulated discharge is approximately 513 m^3/s and, downstream of Sobradinho reservoir, the regulated discharge is approximately 1815 m^3/s.

These numbers were adopted by the SFRBC with recommendation for deeper studies and understanding by all parties involved, in order to make their confirmation or alteration

Table 1. Water availability at São Francisco river basin

Physiographic region	Discharge (m³/s)			Availability (m³/s)	
	Natural average	95% duration	Regulated	Surface water[a]	Ground water[b]
Upper	1189	289	513	622	29
Middle	2708	819	513	1160	294
Sub-middle	2812	842	1815	1838	313
Lower	2850	854	1815	1849	318

Notes: [a]Regulated discharge plus 95% duration incremental discharge.
[b]20% of renewable reserves.

possible when the plan is revised in the future. According to the SFRBC, water availability in the São Francisco river mouth corresponds to a discharge of 1849 m³/s. This number is a result of the maximum regulated discharge in Sobradinho reservoir, added to the 95% duration incremental discharge between Sobradinho reservoir and the river mouth.

Table 1 presents a summary of the water availability at São Francisco river basin. The total water availability is not equal to the sum of the superficial and underground availabilities, as the availability of groundwater accounts for part of the river base run-off.

With regard to the qualitative availability and taking into account the surface water availability and the mean discharge, São Francisco river meets Class 2 conditions (CONAMA, 2005). This means, among other standards, that BOD concentration should be less than 5 mg/L and that DO concentration should be higher than 5 mg/L. In general, with regard to the tributaries of the downstream, middle and sub-middle São Francisco river sub-basins, the Class 2 organic load absorption problem is mainly linked to the low discharges of the water bodies, while in the upstream São Francisco river sub-basins, the problem is mainly linked to the high organic load resulting from high population concentration and industrial activity.

Average demand for water resources at São Francisco river basin in the reference year of 2000 is 168 m³/s. Consumptive use is estimated at 108 m³/s, and the return discharge is 60 m³/s.

The implementation of the water resource management instruments should be in tune with the outlined scenario of water availability and the demand for water resources in the basin. Based on this, the following aspects should be highlighted:

- Water availability in the basin, represented by Sobradinho reservoir regulated discharge, is 1815 m³/s (1931–2001).
- Water availability reduction over the time, due to the increase in the consumptive uses in the basin, today, is approximately 100 m³/s.
- A trade-off between the increase in the consumptive use in the basin and the loss of power generation establishes the basis for a process of negotiation and harmonization of interests that extrapolate the limits of the basin and the mandate of the SFRBC.

Multiple Water Uses

The diversity of water use is a characteristic of São Francisco river basin, and this makes the concept of multiple and concurrent uses a concrete aspect in water management in this basin. The mean coverage level per water supply system in the cities is high (94.8%), higher than the national average (89.1%), but there are municipalities with very poor service levels, particularly in the semi-arid region. The coverage level per sewage system (62.0%) is also higher than the national average (53.8%), although the use of water for the dilution of untreated sewage is an important environmental degradation element in the basin. The collection rate of solid waste is 88.6%, lower than the national average (91.1%), and the final disposition is done inadequately in 93% of the municipalities. The goal of the plan is to have sanitation services universalized, including solid wastes. It also underlines the importance of both service management and social participation.

In 2003, irrigation benefited 342 712 ha and, according to assessments jointly done by power generation companies and CODEVASF, it could be pushed up to the 800 000 ha limit with no conflicts regarding multiple uses. The limit of water use for irrigation has an impact on the irrigated area, depending on the adopted technology and management. Hydropower in the basin is strategic and decisive for supplying power to the north-east region. It contributes substantially to the total national production (approximately 17%). The increased multiple uses upstream result in reduction of the water availability for power production, and this implies the need to establish a trade-off between the increase in the basin consumptive use and the loss of power generation in a planned and negotiated way.

Navigation, which is a traditional activity in São Francisco river and in some of its tributaries, can be potentially developed. The following aspects should be considered: living along with a river in the plains, ensure minimal drought, and fulfil the reservoir operating rules. Significantly prone to fishery and aquiculture in the basin, this potential should be encouraged through appropriate techniques focusing regional socio-economic development and environmental conservation.

The guarantee of minimal discharges for ecosystem maintenance and preservation of the aquatic biodiversity need detailed studies in order to be better defined. Water-related tourism and recreation activities present a potential to be explored in the short term, with a significant impact on the various regions of the basin, as long as their development is linked to the concept of environmental sustainability.

In the basin as a whole, the status of water availability is adequate in meeting present and future multiple uses of the basin. However, some specific areas of conflict can be identified in the following sub-basins of the Velhas, Paraopeba, Alto Preto, Alto Grande, Verde Grande, Salitre rivers. Generally, these conflicts involve: irrigated agriculture; power generation (dam construction and reservoir operation); navigation; supply for human consumption; dilution of urban, industrial and mining-related effluents; and ecosystem maintenance.

Development Scenarios

In order to design scenarios aimed at the allocation of water in the basin, for a baseline the Plan used the scenarios adopted in the study designed for the National Electric Power Operator (NEPO). One of them is tendencial (used by NEPO). Another one is normative, and considers the forecasts of the federal government reflected in the development axes and in the approved long-term federal planning. A third scenario, named optimistic, incorporates the expectation to go beyond the goals proposed for the normative scenario.

Table 2. Scenarios and growth rates for the period 2004–2013

Development scenarios		Growth rates	Effective consumptive use in 2004 m^3/s	Forecast of effective consumptive use in m^3/s
Tendential	Grows according to rates similar to current ones	1.9% per yr	90.9	107.9
Normative	Grows according to rates similar to the Pluri-Annual Plan's (PPA)	6.5% per yr	90.9	134.9 w/o interbasin transfer 160.4 with interbasin transfer*
Optimistic	Grows according to rates higher than the similar to the Pluri-Annual Plan's (PPA)	8.9% per yr	90.9	169.6 w/o interbasin transfer 195.1 with interbasin transfer[a]

Note: [a]The interbasin transfer discharge corresponds to $26.5\,m^3$ for water supply.

In the basin, it was estimated that a number of large-sized enterprises should be carried out in various development phases. In the proposal submitted in the Plan, these enterprises were considered in both the normative and in the optimistic scenarios, the only difference was the allocated discharge. After a review by the TSG, two new forecasts were added, disregarding the discharge related to the São Francisco river water interbasin transfer to the north-northeast region, in both above mentioned scenarios. Table 2 presents a summary of the development scenarios taking the proposed changes into account.

Water Allocation, Classification of Water Bodies, Integrated Control and Charging for the Use of Water Resources

The instruments related to water allocation, classification of water bodies, control and charging for the use of water resources were fully considered in the Plan. In this context, the proposal for the Water Pact in the basin stood out for the sustainable use of the water resources. In the São Francisco river basin there are federal and state rivers. One of the major management challenges in this basin is to establish a minimally harmonized environment with regard to regulations and procedures for permits, charges and control. The proposed Water Pact, to materialize in an Integrated Management Agreement, reflects precisely the search for the solution of this challenge, through an agreement involving the federal government, the Federation Units, and the Basin Committee. In this agreement, each one of the six states and the federal district should commit to a minimal condition of quality and quantity for the delivery of the water of the tributaries under their jurisdiction, into the federal São Francisco river.

Integrated Management Agreement

In order to establish a harmonized environment with regard to laws, regulations and procedures that allows the implementation of water resource management instruments at São Francisco river basin, an agreement was first proposed involving the federal government, the Federation Units, and the Basin Committee, a Pact for Water, to materialize in an Integrated Management Agreement (Figure 2).

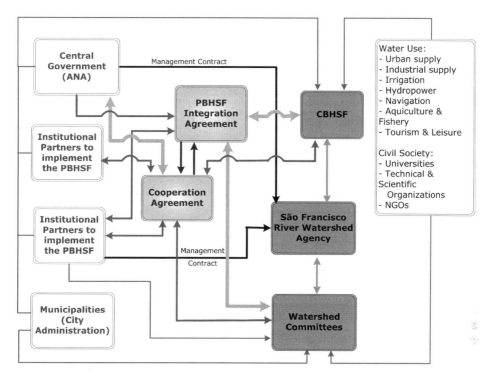

Figure 2. Agents involved in the Integrated Management Agreement

In its initial version, the São Francisco river basin is divided into six regions (Figure 3). Each state would undertake the management of its the water resources, committing to ANA and to the remaining states, to ensure the minimal conditions of quality and quantity agreed upon in the Agreement.

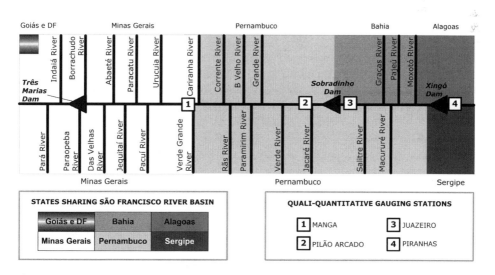

Figure 3. São Francisco river basin division proposed for the management of the water resources

In the version approved by the Committee, the signature of the first Integrated Management Agreement was confirmed as the initial phase for the construction of the Water Pact in the basin, but a choice was made to revise the commitment that guaranteed minimal delivery discharges and water quality. In this context, the following initial objectives were defined to be met in the short run: (a) to promote and implement the registry of the water resources in the basin; (b) to revise previously issued water permits; (c) to implement the registry of users of the water resources in the basin; and (d) to implement a computerized system to manage the water resources, integrating the committees and the managing agencies. The outcomes achieved should feed the new discussion and negotiation of the water allocation proposal in the basin.

Water Allocation

After the simulations for water allocation under the different scenarios, a discharge of $380\,m^3/s$ was proposed as the maximum allocation discharge in the basin that allows meeting the consumptive needs estimated in all scenarios in a relatively comfortable fashion, of which $330\,m^3/s$ would be allocated in the sub-basin defined upstream of Piranhas gauging station, located right downstream of Xingó dam.

Simulation runs allowed the following conclusions with respect to water allocation in the basin:

- The basin can easily afford a $380\,m^3/s$ availability to be split among all current and future consumptive demands within the period 2004–13. This availability will comfortably meet any demands forecast for the three adopted scenarios over the upcoming 10 years.
- The 2013 demand might also be met, in the three scenarios, at a lower level of allocation. However, the states could face little flexibility in order to use water for future requirements in more critical points.
- The $380\ m^3/s$ allocation allows meeting the estimated consumptive use in 2025, but it is recommended that future Plan reviews do harmonize the allocated discharges with the actually observed consumptive use over the years, by assessing the opportunities to add allocated discharges and the subsequent increase in affordability regarding consumptive use.
- The currently granted consumptive uses can be met at an allocated discharge of $380\,m^3/s$. However, as the current $90.9\,m^3/s$ consumptive use represents only 27% of the granted consumptive use, a review in the procedures and examination of the permit requests is recommended, as well as a gradual revision the previously issued permits.
- Restrictions regarding minimal discharges to meet environmental requirements and navigation conditions are also comfortably met.
- The discharges left from the $380\ m^3/s$ allocation allow the goals of classification of the water courses to be met, as long as the sewage treatment level in the basin is increased.
- The allocation consumptive use proposal of $330\,m^3/s$, all the way to Xingó, indicates that this discharge will not be available for electric power generation. The power that will not be generated should be obtained by other means, and this will imply costs to replace the source of power generation. Considering that there

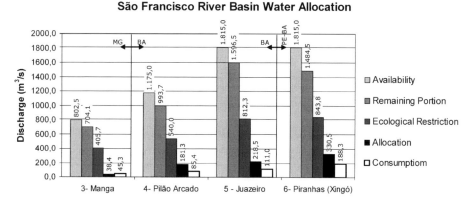

Figure 4. Water allocation at different checkpoints

is already an estimated average consumptive use of 90.9 m³/s in the basin, the impact resulting from a consumptive use increase of up to 330 m³/s is computed as an average of 548 MW.

The water allocation proposal implies defining maximum allocated discharges for consumptive use in each sub-basin and reaches of São Francisco river, and defining the remaining minimal discharges to be monitored in the checkpoints of the main river (Figure 4).

Figure 4 shows that the remaining discharges are far higher than the restricted discharges at the checkpoints. There is still a considerable surplus between the maximum allocated discharge and the consumptive use discharge. In order to effectively meet the checkpoint conditions, appropriate reservoir operation rules should be defined. In addition, it is paramount to set up a Monitoring Technical Group that will assess consumptive use progress and follow up the achievement of the agreed conditions at the checkpoints.

In revising the originally proposed figures for the Plan, the following aspects were considered by the TSG:

- A daily average discharge of 1300 m³/s was adopted as the minimal ecological discharge for São Francisco river mouth, while the annual average discharge of 1500 m³/s was adopted as the remaining discharge in the river mouth. These figures were confirmed by SFRBC Deliberation No. 08, until the Plan is again revised.
- The operation of the electric power sector reservoir is a complex process and is subject to contingencies that could affect the effluent discharges, thus reducing water availability in the channel.
- Determining water availability is subject to inaccuracies and approximations that are inherent to the evaluation of variables representing natural phenomena.
- Ensuring basin sustainability requires the establishment of a strategic reserve both to cope with critical hydrological events, and to enable new enterprises not foreseen in the Plan timeline.

Due to the aspects above, the provisionally adopted maximum consumptive use to be allocated in the basin was 360 m³/s, prescribed in SFRBC Deliberation No. 08.

This number, although lower than the originally proposed $380\,\text{m}^3/\text{s}$, also allows comfortably meeting the demand related to all effective consumptive uses planned for the three scenarios (Table 2).

The water allocation proposal, in addition to defining the maximum discharge to be allocated, also implies defining the minimal remaining discharges. As noted above, rules were proposed for the discharge delivery into the tributary rivers, and for the monitoring of these discharges in checkpoints along São Francisco river. The remaining discharges were confirmed to be higher than both the consumptive use discharges and the restriction discharges at those points. Aiming to consolidate water allocation in the basin, a spatial allocation proposal was prepared in order to be negotiated by the states. The agreed figures, after the negotiation, would be assigned to the Integrated Management Agreement, in order to guide the revision work of the permit requests submitted to each state-level management agency.

SFRBC Deliberation No. 10 refers to the follow up of the agreed conditions at the checkpoints, but it does not define those points or how the conditions to be agreed upon or defined should be observed. In fact, by understanding that the rules and figures proposed require more examination and discussion for approval, the SFRBC defined in Deliberation No. 08 that the minimal delivery discharges in the outlet of São Francisco river tributaries should be discussed during the following plan update, being negotiated during the process of construction of the Pact for Water, concurrent to the permit and registration process revisions. Therefore, according to this deliberation, the minimal delivery discharges in the river mouth of São Francisco river tributaries should be those resulting from the adoption of the permit criteria already practiced by the states, as the new rules are not yet defined.

Therefore, the approved basin water allocation proposal is restricted to indicating the maximum discharge to be allocated and to preserving the criteria existing in the states, postponing the definition of the rules for minimal delivery discharges and for the spatial distribution of the maximum allocated discharge, postponing the consolidation of the Pact for Water and the Basin Integrated Management Agreement.

General Guidelines for Water Use Permit

The permits existing in the basin perennial rivers, issued by the states and by the federal government based on the maximum basin discharges, total $582\,\text{m}^3/\text{s}$, and translated into granted consumptive use, represent approximately $335\,\text{m}^3/\text{s}$. With a current average consumptive use of approximately 27% of the granted volume, a proposal was made so that the technical review procedures, the permit criteria, and the existing permits were revised by the states and, more importantly, by the federal government.

As far as the negotiated permit revision is concerned, SFRBC Deliberation No. 09 recommends the establishment of a Compensatory Chamber consisting of representatives of the Federation Units that are part of the basin, ANA, and to consider the following premises: (a) the process should not imply suspending or cancelling the permits, but to revise the maximum discharges estimated for 2013; and (b) the consumptive use of the new permits added to the existing permits should not exceed the figures for maximum discharge to be allocated defined in the Plan. This deliberation also includes the minimal technical criteria to be adopted in the process of revision, and recommendations with regard to the registration of users and the need to ensure an integrated review of the environmental impact produced by the enterprises.

As a result of the TSG's effort to examine the preliminary proposal, in addition to the guidelines and criteria of permit revision, the Plan also received some criteria, limits and priorities for water use permit, consolidated in SFRBC Deliberation No.11. A summary of the most important proposed aspects is as follows:

- An examination of the permit requests should be guided by the following points: (a) water use efficiency; (b) proof of technical, socio-economic and environmental feasibility; and (c) harmony of multiple uses.
- The permit requests for human consumption and animal use should be a priority, such as predicted under Law 9433/97. A limit of water loss (30% for projects not yet implemented), and a commitment with the appropriate destination of the effluents produced should be observed. With regard to the already implemented projects, a period of five years shall be established for them to comply with the quantities and goals. In case of project expansion, whatever is agreed between the entrepreneur and the granting agency should be fulfilled.
- The basin's agricultural aptitude should be met, provided the requirements for examining the permits are observed.
- With regard to requests submitted to the permit officials, large-sized enterprises (maximum catchment discharges equal to or greater than 5 m^3/s) should report to the SFRBC for their information.
- Water catchment and bypasses of up to 4.0 L/s of installed capacity in the São Francisco river channel are regarded to have little importance and, despite being subject to the registration process, they do not depend on the permit. This limit should be reassessed when the sum of the installed capacity exceeds the amount corresponding to 0.5% of the long-term natural mean discharge in any stretch of São Francisco river.

Classification of Water Bodies

The classification proposal presented in Figure 5 was based on all studies conducted for the rivers of the São Francisco river basin, considering the diagnosis and their predominant uses, as well as the current situation of water quality. These concepts were approved in SFRBC Deliberation No. 12.

With regard to the approved proposal, two aspects should be emphasized because they relate to future activities:

- The technical discussions on classification should continue in order to improve the current stage of knowledge, mainly on studies related to intermittent basin rivers, the information available on which are not yet sufficient to support the classification of these rivers.
- The consolidation of the approved classification depends on SFRBC's actions and their respective Water Agency together with the public institutions, so that measures aiming to meet the water quality goals are implemented. These measures make up the Classification Consolidation Plan.

Despite approving the concepts of the classification proposal, SFRBC Deliberation No. 12 did not specifically handle the proposal, based on a list attached to the Plan where classes are assigned to the quality of 96 reaches of water bodies. In order to be legally valid, this

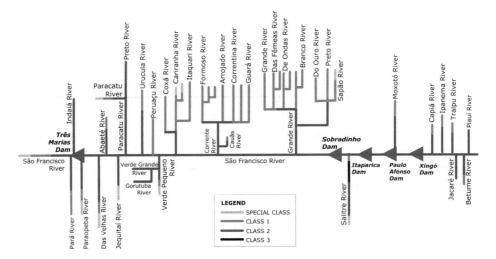

Figure 5. Proposal for São Francisco river basin water-body classification

classification proposal should be submitted by the Committee to the CNRH for the federal water bodies, and to the Water Resource State Councils for the state water bodies to be examined and approved.

Guidelines and Criteria for Charging for the Use of Water Resources

Considering the size and complexity of São Francisco river basin, the main premise of the charging methodology discussed in the Plan is its operational and conceptual simplicity, as: (a) it can be immediately adopted; (b) it diminishes the risk of significant economic impact on paying users; (c) it indicates the importance of water resource rational use as far as water quantity and quality are concerned; and (d) it facilitates official control by the agencies in charge, and user control by their own peers and by the basin Water Agency.

These points were made during SFRBC Deliberation No. 16, which outlines that the Committee should conduct technical studies in the short term by means of the Permit and Charging Technical Chamber supported by ANA, aimed at establishing the charging mechanisms for the use of the basin water resources, and the prices to be charged.

Before becoming effective, it is important that the charging is preceded by a comprehensive process of negotiation with the basin stakeholders, and that the implementation is conditioned to an advanced guarantee of full investment of the collected funds in actions for the basin itself, pursuant Art. 22 of Law No. 9433/97.

Integrated Control and Monitoring of Water Resource Use

Control and monitoring should guarantee the multiple uses and the appropriate meeting of the demands and priorities of use established in the Plan, and will result from an arrangement between the federal government, the states and the federal district. Based on the proposal submitted in the Plan and in SFRBC Deliberation No. 13, both of which are complementary, a summary is given below of the main premises and actions recommended for integrated control and monitoring of the use of water resources at São Francisco river basin.

- Control actions should have a mentoring, educational and preventive role, but this should not preclude the enforcement of the penalties, when infractions occur.
- Federal and state control agencies will always act as partners, joining efforts to adopt administrative criteria and routines aiming to harmonize procedures taking into account the whole basin, to be edited in a document that expresses this consensus of the involved parties.
- The water quality monitoring network should be expanded, prioritizing the establishment of stations representing the contributions of the main tributaries, integrated to the monitoring of the basin water volume.
- A reassessment should be made of the strategic definition of the priority areas for control and monitoring, contained in the Plan.
- The criteria and figures of ecological discharges for the basin rivers, regulated reaches and river mouth, should be determined from the promotion of studies and implementation of a basic network for a self-developed methodology.
- The agencies that make up the National Water Resource Management System that covers the basin should bring their efforts together in order to design and operate a Contingency Plan based on the mapping of points and situations of potential ecological accident risk.

In the preliminary basin water allocation proposal, the importance of forming a Monitoring Technical Group was emphasized, as this TG is a key component to assess the evolution of water consumption, and follow the enforcement of the conditions established in the Pact for Water. In SFRBC Deliberation No. 10, the implementation of an integrated system to inspect and monitor the use of basin water resources with the very same attributes is recommended, postponing the decision concerning its operational aspects. The same deliberation proposes the establishment of a Technical Chamber of Research, Technology, Information and Monitoring, within the Committee.

Strategy for Hydro-environmental Revitalization, Recovery and Conservation, and Investment Programmes

From the identification of the demands, and an intensive participatory process, the interventions to be included in the plan and the relevant investments were selected, organized in a physical-financial schedule, indicating the possible sources of funds. This selection led to a comparison between the existing reality and the desired reality, as well as to the definition of objectives and goals that, in turn, depend on the capacity of the society and of the SFRBC to promote the changes deemed necessary.

The reference base to structure this item comes from SFRBC Deliberation No. 03 that defined components and activities that should make up the Plan, according to the content of Art. 4:

the environmental revitalization of São Francisco River Basin, considered as the hydro-environmental recovery of the basin, consists of a group of measures and management actions, projects, services and works, that make up a planned, integrated and whole project within the reach of the basin to be developed and implemented by local governments, by the federal district, by the state, by the federal government, by the private sector and by the civil organized society, aiming to recover both quality and quantity of surface and groundwater, taking into account the guarantee of multiple uses, the preservation and recovery of the basin biodiversity.

Selected Interventions

The following aspects, among others, guided the selection of the appropriate interventions for the hydro-environmental recovery and conservation of the basin:

- the level of complexity and heterogeneity of the basin, characterized by different purposes and objectives established for the Plan, and by a variety of established demands;
- matching the identified needs with the availability of resources, their flow over the time, and the ability to use them efficiently, which has required choosing some actions rather than others;
- the definition of a group of criteria to be met by the selected interventions, the process of which should be periodically conducted in order to comply with the transformations observed in the basin and in the institutional framework; and
- the checking of the actions contained in the São Francisco River Basin Revitalization Programme, designed by a Technical Group made up of representatives of institutions such as: MMA, IBAMA and ANA, coordinated by SECEX-MMA. This programme was established by the President's Decree of 5 June 2001, and included in the PPA 2004–2007.

The selected interventions were organized at four levels—components, activities, actions and individual interventions—with increasing degrees of disaggregation, capable of successfully meeting the needs of the Plan (Figure 6).

The first level is the most comprehensive of all with a higher degree of aggregation and lower spatial resolution, covering the basin as a whole, and corresponding to the components already consolidated in the first moves towards the design of the Plan, starting from SFRBC Resolution No. 03, including a component dedicated to the semi-arid region. In SFRBC Deliberation No. 14, approved by the Committee, the Plan components were presented as follows:

- *Component I*: Implementation of both the Integrated Water Resource Management System (SIGRHI) and the Basin Plan.
- *Component II*: Sustainable Use of Water Resources, Hydro-environmental Protection and Recovery of the Basin.
- *Component III*: Water Resource Services and Works and Use of Basin Land.
- *Component IV*: Basin Environmental Quality and Sanitation.
- *Component V*: Water Sustainability of the Basin Semi-Arid Region.

In spite of the slight nomenclature difference of the components with regard to the original proposal of the Plan, the content and objectives of the activities and actions expected are basically the same.

The five components of the Plan are divided into a total of 18 activities and 48 actions. The revision prepared by the TSG expanded that number to 29 activities and 139 actions split into the very same five components, mainly based on the results of the public

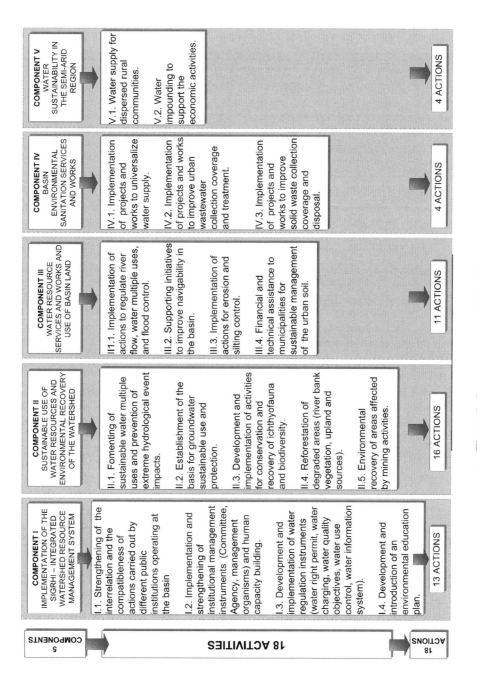

Figure 6. Structure proposed in the Plan

discussion rounds of the Regional Advisory Chambers held in the four physiographical regions of the basin (upstream, middle, sub-middle and downstream São Francisco River), and on the selection criteria of the interventions. The new proposed number of activities and actions (29 and 139, respectively) was approved in SFRBC Deliberation No. 14, although an estimate of the investments related to these new activities and actions has not been made.

Investment Programme

The Plan Support Technical Studies allowed to estimate the investments in approximately R$5.2 billion, shared by the five proposed components for the implementation of the 18 activities and 48 actions initially planned (Figure 7).

The list of investments was adopted by the Committee in accordance with SFRBC Deliberation No. 15 that foresees, however, that the Investment Programme undergoes a reassessment and a revision in the short term. The strategies and procedures to be adopted in the revision process are outlined in SFRBC Deliberation No. 15, including the following objectives:

- detailed reporting of the actions that are part of the group of interventions, in order to define the specific interventions for the period 2006–13;
- hierarchical definition of the specific actions and interventions, and their spatial distribution in basins of tributary rivers and reaches of São Francisco river;
- definition of strategies and implementation of interventions and goals to be met in terms of hydro-environmental recovery and conservation;
- definition of the participating organizations and potential executing agencies and institutions; and
- appraisal of costs, sources of funds and deadlines for each specific intervention.

With regard to the Plan's initially proposed 18 activities and 48 actions, a substantial number of these objectives have already been met, including the distribution of the necessary funds to the various actions, activities and components; the phases when these funds were used; and the possible origins of financial resources.

Concerning the distribution of the necessary resources, the actions and activities of Component IV (environmental sanitation) accounted for 84.1% of the R$5.2 billion

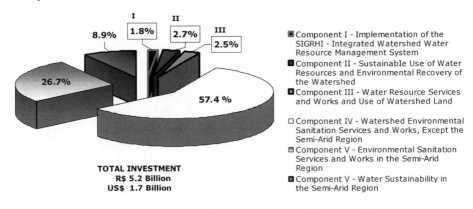

Figure 7. Investment allocation for the Plan according to the Components

estimated for the Plan implementation. Considering investments in sanitation and in water sustainability actions (Component V), approximately 35.6% of the total estimated investment will take place in the semi-arid region.

Although requiring less investments (approximately 1.8% of the total), from the point of view of the Plan implementation, the actions regarded as most important are connected to Component I (Implementation of the SIGRHI). SFRBC Deliberation No. 15 also recognizes the importance of Component I when prioritizing the following activities for the period 2004–05: (1) implementation of the studies and actions necessary to designing and constructing the Pact for Water and the partial revision of the Plan; (2) implementation of the activities and actions of the SFRBC, including its institutionalization and the Basin Agency operation; and (3) information and research management.

From the perspective of the use of the financial resources, three different phases were foreseen for the Plan: (1) initial (2004–05), in which the use of the resources will be reduced and implementation efforts will be concentrated on Components I and II; (2) intermediary (2006–09), in which the need for resources will quickly rise until reaching its annual peak, and the implementation effort will be concentrated on structural interventions; and (3) final (2010–13), in which the demand for financial resources will slightly decline over the years.

The main sources of funds predicted for the Plan are the overall federal government budget and state funds, Hydropower Compensation to States and Municipalities, funds from public service utility companies, revenue from the use of water, and international financing. A portion of the funds from these sources is guaranteed to the basin, but a substantial portion of these resources has to be negotiated. Therefore, two actions proposed in SFRBC Deliberation No. 14 gained importance:

- arrangements with states that share the basin, aimed at the participation of the SFRBC in the definition of priorities and lines of action of their basin revitalization programmes; and
- arrangements with CHESF and CEMIG aimed at the participation of the SFRBC in the definition of priorities and lines of action towards revitalization, creating lines of credit for research and development, and in joint actions of support to the Regional Advisory Chambers.

Strategies for the Implementation of the SFRBP

In order to be implemented, it is crucial that the Plan explicitly incorporates the main economic-financial and political-institutional conditioning factors in its decision-making process, to ensure the following aspects: (1) the resources, especially those of financial nature, to carry out the Plan; (2) the internal organization and operation of the SFRBC and of the Agency to be established; (3) both operating and water-related sustainability of the interventions planned for the basin; and (4) decision makers' commitment to the Plan, which implies representativeness of the SFRBC in the basin by means of the support and mobilization of the society, and in the gathering of support from the basin external sectors.

These requirements outline the implementation fronts of the SFRBP, each one of them to be provided with a specific strategic treatment, as follows:

- *Economic*: The SFRBC and the Agency will be in charge of negotiating in order to include, in the federal government's and in the states' budgets, funds for the interventions listed in the Plan, as well as their disbursement. The definition of a time schedule for the implementation of the charging for the use of the water resources is also strategic to indicate the endeavour of the basin to implement a water resource management and their commitment to take part in the investment programme, the initial steps of which were defined in SFRBC Deliberation No. 16.

- *Institutional*: The importance of this front reflects the discussion power of the players involved in the implementation of the Plan. In this context, the composition and the implementation of the basin Agency plays a paramount role in order to implement the executive actions relevant to the Committee, to provide technical input for decision making, and monitor the implementation of the Plan and of the Pact for Water, according to the recommendations of SFRBC Deliberation No. 10.
- *Technical*: The Plan's technical consistency and operation, among other principles, should be based on the sustainability of the environment and of the water resources, as well as in the harmony of their multiple uses. Based on these general principles, the Plan also includes specific technical recommendations to the user sectors, discussing their roles and their different forms of contribution to the Plan objectives.
- *Social*: The players who are not directly involved in the execution or implementation of the Plan are important elements for the mobilization of the efforts and expansion of their support base.

Plan Implementation Map

The Plan should be viewed as a map that leads to a specific destination, the overall objectives that guide it. In times of uncertainty, and in an area of a high level of complexity, the choice of the avenues to be explored should make use of a strategy that complies with the reality of each moment.

After the approval of the Plan, it is necessary to be careful and avoid the post-Plan vacuum. For this purpose, two measures should be immediately implemented: (1) the strengthening of the technical office—an embryo of the basin Agency—to support the implementation of the Plan, while the basin Agency is not completely implemented; and (2) the effective establishment of the basin Agency, that demands some time to be completed, as it requires a number of events to be carried out and the fulfilment of several legal measures.

Plan activities should be started and conducted as concurrently as possible, although adopting different efforts and paces to the various actions, following the previously mentioned adaptive strategy. In this sense, some steps could be identified:

- A set of interventions of easier implementation should be chosen for the first years, especially those that demand less resources or those that face less conjunctural resistance to funding and releasing of the resources. It is convenient to elaborate on the structural interventions integrating this group, making up a project bank to increase their appeal and eligibility for financing by the agencies in charge of releasing the resources.
- The first result of the Plan should be an Integrated Management Agreement to be signed by the federal government and by the states that share the basin, supported

by the SFRBC, endorsing the water allocation criteria listed in the Plan, and delegating to the states, whenever is the case, their adoption in their territories. This Agreement will represent a clear demonstration of value assigned by these Federation Units to the basin, as well as it will be, at the same time, a demonstration of commitment and indication of openness to federal and state investments.

- The first initiative of the Committee after the approval of the Plan and after the Integration Agreement is signed should be to broadly publicize it in the basin, as well as to expand its supporting base. Meetings and presentations should be scheduled to (1) identify opportunities; (2) obtain support; (3) establish commitments to the Plan and consensus regarding the first interventions; and (4) implement the institutional arrangement.

- As a naturally deepened discussion during the meetings and presentations of the Plan, the SFRBC should strive to include the planned interventions, particularly the structural ones, in the federal government's and in the states' budgets. This is an effort to raise awareness, convince, mobilize and join people, and is typical of the role of the Committee that will be supported by the Technical Office and, later on, by the Basin Agency.

- After overcoming the budget approval phase, the tone of these arrangements should turn towards ensuring the disbursement of the funds listed in the budget line items to complete the interventions indicated in the Plan. The economic and political-institutional implementation front will last permanently during the life of the Plan, renewing its objectives every year. It will have the support of the contact network formed during the meeting campaigns and presentations of the Plan, and will benefit of the capacity of public mobilization developed by the SFRBC.

- Biennial reassessments of the progress achieved should be carried out. At this time, priorities will be chosen for the following period based on the results achieved up to that point.

Component I is the key component of the Plan because it involves the implementation of the Integrated Basin Water Resource Management System, SIGRHI. As it is not a structural component, it also consumes less resources that presumably should first come from the governments (federal, state and federal district), from the GEF and, thenceforward, from the charging for the use of water. The possibility of implementing each action listed under Component I will depend on the ability to discuss and negotiate, developed during the implementation of the Plan.

Component II refers to the sustainable use of the water resources and basin environmental rehabilitation. This includes activities and actions materialized by means of studies, guidelines designing, promotion (of use, of activities), support to actions of water use rationalization and conservation, and recovery of biodiversity and conflict management. Therefore, the interventions that are part of Component II are an extension of Component I and broaden their effects. Cooperative management is the strategic approach to be used for this component. The Committee will work together with the sectors in charge or affected, to recover, prevent or conserve, and the SFRBC is responsible for monitoring the studies, supporting and promoting the actions put forward by third parties via cooperation agreements.

These actions can be financed by the executing agencies, NGOs, national and international promotion agencies, and by the federal government and states' budgets. The first Plan dissemination actions might be useful to identify opportunities to obtain financing for the actions of this component. From the point of view of an implementation schedule, Component II should be started after the presentation of the Plan is completed throughout the basin. However, it could start at the same time as Component I. It is important to remember that the actions that make it up are long-term actions, and they should get along until the end of the Plan, and that their pace and progress should be adjusted to the political-financial conditioning factors.

Components III (Water resource services and works, and land use), IV (Environmental sanitation services and works), and V (Water sustainability of the semi-arid region) are of a structural nature, demanding more resources, especially in the environmental sanitation sector. Their implementation will require significant efforts to identify, assign and release the budget resources and the exercise of an adaptive strategy that brings together the perception of opportunities with flexible options and decision-making swiftness. Having a set of pre-arranged interventions sorted out in terms of projects and costs that can meet the requirements of the financing agents, reassessed every two years, will be an absolutely essential tactic for the achievement of some success in this struggle for resources.

As said previously, it is essential that the water resource management developed during the implementation of Component I is complemented, within the first two years, by small structural interventions that are capable of demonstrating the Plan's potential, and give the SFRBP visibility in the basin. These actions should leap in terms of pace and speed in the third year, when more resource funding is expected as a result of discussions, negotiations and contacts established over the first two years.

The most significant source of funds of the interventions integrating Components III, IV and V are the budgets by the federal government and the states. However, in order to be incorporated by these budgets, they must have been accepted in their relevant PPAs. This indicates that the SFRBC should forcefully strive for resources and, in order to do that, it will need significant discussion and negotiation power.

From what has been shown, it is important that the SFRBC has a project portfolio available to give it the necessary flexibility to adjust to the reality it will face during the implementation of the Plan, so it can take the most out of these conditions. It should then develop the ability to conceive different tactics of project financing, such as: (1) the use of local counterparts; (2) the ability to set up competitive funds and support the financing of APL's that make rational use of the water resources; (3) the promotion of micro-regional arrangements (such as the Inter-Municipal Cooperation, and the use of resources for hydropower compensation), among others.

Conclusions

The process of design and approval of the São Francisco River Basin Ten-Year Plan, which translates a significant progress in water resource management and in the effort to develop the basin, could be summarized, in simple terms, in the following steps: (1) preliminary draft of the Plan designed by the GTT, submitted for examination on 30 April 2004; (2) Plan revised by the TSG, and discussion of the main aspects related to water allocation and basin revitalization, in public consultations, in May–June 2004; and (3) discussion and approval of Deliberations Nos. 07 to 17, prepared based on GTT's and

TSG's documents, during the III SFRBC Plenary Meeting, held in the period 28–31 July 2004.

The Plan was originally structured into four modules containing, respectively, the Executive Summary; the basin diagnosis; the allocation and implementation proposal of the water resource management instruments; and the investment programme for the revitalization of the basin, which were all discussed and consolidated until the Plan was approved by the Committee.

The basin diagnosis was prepared on the basis of existing documents, such as the Strategic Actions Programme (SAP) and the Basin Analytical Diagnosis (BAD), as well as on the Plan Support Technical Studies developed within the National Water Agency (ANA) supported by consultants. It presents a rich overview of updated, consolidated and reviewed information about the basin, being an essential element for the remaining phases of the Plan, and also a necessary starting point to new studies and future updates.

With regard to water allocation, the fact that the Plan has only approved the maximum discharge of consumptive use to be allocated, although still at a provisional basis, delays the signing of the Pact for Water and, consequently, of the Integrated Management Agreement, postponing the definition of the rules for minimal delivery discharges, the spatial distribution of the allocated discharge, and the implementation of the agreed conditions.

As a result, even if significant progress has been achieved in other aspects, such as the classification of water bodies, the lack of definition concerning water allocation affects the performance of the whole set of water resource management instruments, as far as their integration is concerned. A number of activities to be completed in the short term should contribute to the new discussion and negotiation of the proposal of water allocation in the basin in order to definitely implement the Integrated Management Agreement for the sustainable use of the water resources.

The activities foreseen in the investment programme for the revitalization of the basin should be started and conducted as concurrently as possible, although adopting different highlights and paces in the several actions, and the Committee should make use of a range of strategies that provide it with flexibility and the ability to adapt to the reality of each moment. In spite of the consensus concerning the importance of basin hydro-environmental conservation and recovery, some issues are still open because several proposals have been added to the original programme without estimating their required investments. Thus, the investment programme will also be subject to reassessments and revisions, so that a consensus can be established with regard to the group of prioritized interventions. Therefore, the post-Plan agenda is full of studies, revisions and adequacies so that the Plan can effectively meet its goals.

From this point of view, it is important to recall the concept of a plan, as a process—an organic and dynamic element that guides decision making in seeking pre-established objectives. The constant attention, perception, interactions and consolidation of the opportunities and materialization of the Plan, by means of political-institutional negotiations and participatory management, should be its most important strategy of implementation, follow-up, monitoring and revision.

Therefore, the Plan is an initial landmark. In short, seeking basin sustainable development, which is the main focus of the Plan, is principally a process of activation and channelling of the social forces, an exercise of initiative and creativity, and an improvement of the cooperation and interaction skills of the different players who live in the basin.

References

Biswas, A. K. (1978) Water development and environment, *Water International*, 3(2), pp. 15–20.

CONAMA (2005) Resolução 357, que dispõe sobre a classificação e diretrizes ambientais para o enquadramento dos corpos de água, bem como estabelece condições e padrões de lançamento de efluentes.

Global Environmental Facility (2000) *Transboundary Diagnostic Analysis Manual* (Washington DC: GEF).

GWP (Global Water Partnership) (2003) *Integrated Water Management Toolbox Version 2.0* (Stockholm: GWP Secretariat).

The Dilemma of Water Management 'Regionalization' in Mexico under Centralized Resource Allocation

CHRISTOPHER A. SCOTT & JEFF M. BANISTER

Introduction

Globally, Integrated Water Resources Management (IWRM) emerged as a conceptual approach to release water resources planning (and to a lesser extent its development and management) from the vice-grip enforced, indeed required, by 'hydraulic despotism' that prevailed for much of the 20th century. However, taking the next step of implementing integrated plans was a leap of faith that banked heavily and in many cases naively on the conformity of large water resources bureaucracies to open themselves to integration. This entailed transparency, accountability and dialogue in political and public circles in which conventional water managers found themselves ill-equipped to defend the central tenets of their profession, virtually exclusively, engineering. They saw their power base being eroded in the integration process. In countries with large water bureaucracies, the pace of 'integration' (immersing water management and managers in broader institutional, political, and public decision-making processes) has been excruciatingly slow (Shah *et al.*, 2004).

IWRM was conceived on the drawing boards of global development paradigm designers (read Stockholm, Dublin, Rio, the World Water Forum process, Bonn, WSD, etc.), concerned with water sector development as a means to pursue broad social and economic growth (Varady & Meehan, 2006). However, the design of paradigms was at least one step removed from water resource implementation and management in real-world situations. As a result of glossing over the full complexities and intricacies of water management in operational terms, there never has been a clear, concise IWRM roadmap; indeed, the planning process is incapable of capturing such complexity.

It is necessary to provide a broad characterization of the functional attributes of water resources planning and management. There is a structural dualism inherent in the professional roles played, on the one hand, by public policy makers and planners whose objectives are centred on social welfare with a distinct bias on the electoral balance sheet, and on the other by technocrats charged with efficiency, functionality, and effectiveness, attributes these same technocrats, it might be added, would not be averse to using when characterizing their own trade. As long as water agencies retained a single purpose (or single function) water use mandate, e.g. irrigation or hydropower, but not urban water supply which will be returned to subsequently, there were few barriers separating technocrats from policy makers: an engineer (or other career professional) with the right charisma or 'inside track' faced little difficulty in becoming the head of the irrigation department.

However, the inexorable process of institutional diversification resulting from bureaucratic reform or public sector modernization combined multiple functional attributes into single, invariably larger governmental bureaucracies. For example, this occurred in Mexico with the relocation of irrigation from agriculture to the water resources portfolio of public administration; the Ministry of Water Resources (*Secretaría de Recursos Hidráulicos*, SRH) was a gargantuan bureau with cabinet-level status, created in the 1940s when party and state became fused in large technocratic agencies. In the new context that emerged in the 1990s (water subsumed under the environment ministry), the functional barrier between technocrat and policy maker was heightened, making it increasingly difficult for all but the most savvy and politically connected irrigation engineer to rise beyond the rank of section chief. However, two aspects of this evolution should be borne in mind: (1) the resources at the section chief's disposal (human, financial, and critically, water resources) may not dwindle to the same extent that his decision-making autonomy is curtailed; and (2) the policy maker—invariably a political appointee—is increasingly beholden to the technocrat to generate impact while avoiding negative outcomes.

One salient feature of urban water supply sets this sub-sector apart. The exposure of urban services provision, including *inter alia* water supply and sewerage, to public and media scrutiny has led to the politicization of its administration at least a full generation before irrigation, hydropower and related water services. In addition, the public health emphasis (water supply and sanitation as the principal strategies to reduce infant mortality and the burden of gastro-intestinal disease) of social services investment during the latter half of the 20th century, globally but also in Mexico, translated rather predictably into the Millennium Development Goals on access to water supply and sanitation, etc. The global investment and resulting scrutiny such lofty targets have conferred on local or state WATSAN-type programmes have increasingly removed their administration from the technocrats' ambit.

There is also a core-periphery dimension to urban versus agricultural water management and the degree of scrutiny to which each is subject. Urban areas, and by extension urban services provision, are the centre of political power and electioneering as well as revenue mobilization and expenditure. By comparison, the agricultural hinterland has served urban interests primarily for the provision of cheap food, raw material for urban and peri-urban industry, surplus capital for both the state and agribusiness, and, until recently in Mexico, a secure political base for the dominant political party. With few exceptions, GDP growth has resulted in the depopulation of rural areas, or at least a widening demographic imbalance between city and countryside. Where food production is predicated on major irrigation development, i.e. where agricultural water management retains significant resource investment, the combination of de-emphasized rural hinterland with continued injection of financial resources has created an institutional vacuum readily filled by large landowners and their allied interests. The institutional landscape is made more complex by the systematic evisceration of *ejidos* (roughly, agrarian reform smallholder communities), although the very presence of rural poverty remains a compelling argument for rural programmes.

In this context, there are increasingly vocal calls for a societal dialogue process around water resources management with an IWRM backdrop. Local interests (farmers' organizations, citizens' groups, and non-governmental organizations of all types and ideological affiliations) are mobilizing around access to water, access to water resources decision making and water quality impacts. These civil society organizations are increasingly vocal and globalized, i.e. networked with access to knowledge and advocacy (Ruiz-Marrero, 2005), and see in IWRM a chink in the armour of centralized decision making. What effective impact this has had on public policy and programmes remains a crucial question to be researched.

In official circles in Mexico, 'social (or public) participation' has moved from being something external to any but the most localized water management decision making, to being a process that was recognized but studiously avoided, to today's full-on calls for public debate (many of which are acrimonious, fewer harmonious). The net result of IWRM's chequered history has been acknowledgement by technocrats and decision makers that broader societal goals, participation and dialogue are now features of the landscape. However, despite this agnosticism on the part of technocrats, studies continue to highlight ongoing lack of official consideration for the deeply social, political and cultural nature of the hydrological systems they control (Whiteford & Bernal, 1996; Wilder & Romero Lankao, 2006)

Integrated Water Resources Management: A Brief Repast

Biswas (2004) discusses and energizes the polemic around the concept of IWRM, concluding that despite great promise and fanfare enlisted by IWRM proponents, it remains principally an article of faith. One of the central tenets of IWRM is that the immersion of water management in larger public administration processes will be necessary for improved social welfare, economic development and environmental sustainability. This appears to emerge unscathed from the critiques of IWRM by Dourojeanni (2003); Mollard & Vargas (2004); Biswas *et al.* (2005); Tortajada, 2005; and Tortajada *et al.*, 2005). Historically, as a process, water management has been decoupled from broader societal goals; however, its role in contributing to these goals is conceptually unassailable. Or, to clarify this statement, by institutionalizing and personalizing change agents, water management agencies and decision makers have had limited effectiveness in contributing to social, economic and environmental goals.

Tortajada (2005), here with reference to Mexico, captures IWRM's conceptual underpinning:

> Despite its importance, the construction of hydraulic and related agricultural infrastructure projects alone will not suffice. What is required is a water sector with a management vision that integrates natural and human resources such that resulting projects have lasting, beneficial impacts. (p. 7, translation by authors)

IWRM remains a conceptually attractive proposition. The introduction alluded to the global paradigm designers as the primary proponents of integrated approaches. At the policy level, organizations such as the American Water Resources Association (AWRA) continue to promote IWRM. As recently as February 2005, the AWRA posited four water policy challenges, the first and fourth are particularly notable for the present discussion: (1) promoting more integrated approaches; (2) reconciling the current ad-hoc [US] national water policy; (3) developing collaborative partnerships; and (4) providing information for sound decision making. The call for action and conclusions of the 2nd Water Policy Dialogue underscored the integration message (AWRA, 2005).

However, history is replete with well-intentioned designs relegated to the scrap heap when confronted with insurmountable hurdles, or worse, those that continued to be propped up despite realization of their failures or shortfalls (Scott, 1998). Biswas *et al.* (2005, p. 254) identify IWRM's implementation and institutional impasses as follows:

> ... operationally it has not been possible to identify a water management process that can be planned and implemented in such a way that it becomes inherently integrated however this may be defined, right from its initial planning stage and then to implementation and operational phases ...

In the real world, integrated water resources management, even in a limited sense, becomes difficult to achieve because of extensive turf wars, bureaucratic infighting, and legal regimes (like national constitutions) even within the management process of a single resource like water, let alone in any combined institution covering two or more ministries which have been historic rivals. In addition, the merger of such institutions produces an enormous organization that is neither easy to manage nor control.

Are these simple bottlenecks that, with the appropriate measure of coordination and institutional discipline, can be overcome? Or, to the contrary, are they fatal structural flaws in IWRM's broad sweep? To examine this question IWRM implementation issues will be considered, particularly decentralization in the context of Mexico's river basin regionalization, and then the opinions of the authors will be offered in the conclusions.

Integrated River Basin Management: IWRM's Better Half

IWRM in the Latin American context has been viewed as a coordination, consensus-building and conflict management measure (CEPAL, 1999; CONAGUA, 2001). It is recognized that for effective coordination of functions, decentralization of water management is both inevitable and desirable. Dourojeanni (2003) characterizes administrative systems for water management as those with: (a) a large number of institutions under limited central coordination; (b) a high degree of institutional decentralization of functions; and/or (c) the absolute or complete centralization of authority with limited or non-existent delegation of responsibilities. In relation to the Mexican experience, it is necessary to add the 'and/or' in the last set of observations, based on the perception of the simultaneous (and contradictory) existence of water management decentralization with centralized resource allocation.

This contradiction and the need for separation of water management and allocation functions were identified in the CEPAL (1999) report:

> Although the concrete dynamics of reforms vary between countries, they all point to-wards the possibility of creating future systems that apply the concept of IWRM at the river basin level—with a clear distinction between the responsibilities for water management, on the one hand, and its use, on the other … The cornerstone of such restructuring is the separation of public service provision from supervision and regulation … and water allocation from management. (p. 7; translation by the authors)

The tension between decentralized management, on the one hand, and centralized resource allocation on the other, is hardly surprising when situated in historical and geographical contexts. Mexico's irrigation boom, largely initiated in 1926 with the promulgation of the National Irrigation Law (*Ley sobre Irrigación con Aguas Federales*), marked the beginning of a radical expansion of state authority into the countryside (Aboites, 1998). Here, the Yaqui Valley, in the north-western state of Sonora, is emblematic. The Río Yaqui irrigation district presently comprises some 220 000 ha, mostly in wheat, and incorporates the labour of 20 000 producers, from Yaqui indigenous production societies (*sociedades*), collective *ejidos*, and small private-holders (*pequeños propietarios*), to large-scale agribusiness conglomerates (Wilder & Romero Lankao, 2006, p. 1987). As a result, the history of damming and diverting the Río Yaqui is fraught with ethnic tension and social strife. However, it is also the political landscape in which the contemporary push to re-territorialize access to water resources plays out.

The struggle over land and water in the Yaqui Valley, as elsewhere in Mexico, largely revolved around attempts to harness natural resources and human labour for extra-local ends. From the early 1600s to the mid-1700s, the Jesuits' programme of agricultural development, including the introduction of wheat and livestock into Yaqui indigenous communities, created surpluses for the burgeoning California mission system. After Mexican independence (1821), large estates (*haciendas*) produced staple crops and raised cattle using indigenous labour and Río Yaqui water, while viewing themselves as a force for (European) civilization in the region (Spicer, 1980). However, the most radical transformation of land and water in the valley before the 20th century was set in motion during the second term of President Porfirio Díaz (1884–1911), when the Los Angeles (US)-based Richardson Construction Company was granted concessions to most of the valley's productive resources. Despite its contractual obligation to provide water to Yaqui indigenous communities, Richardson gave preferential treatment to its own colonists, and to those deemed most able to produce a marketable surplus. Thus, in 1912, the Yaquis complained of a 30% reduction in their harvests for lack of irrigation water (Aguilar Camín, 1977, pp. 57–58). Such asymmetry would continue to characterize the distribution of the valley's water resources.

The Mexican Revolution (1910–ca.1920) set in motion events that would lead to the federal government's cancellation of the Richardson concession in 1926. However, at least three basic patterns, laid down during the Richardson era, would remain in place to shape the course of subsequent agricultural development (and recent moves to decentralize management and allocation). First, the Richardson experiment (and others like it in Mexico) had produced a deep conceptual rescaling of irrigation control. Now, officials could think in terms of altering entire river basins. Indeed, according to Aboites (1998, p. 73), this was the practical beginning of the hydrographic basin-cum-management unit, at least in the Mexican context. Second, notwithstanding the revolutionary rhetoric and practice of agrarian reform, private capital would continue to play a dominant role in river basin resource politics, a situation that over time grew increasingly acrimonious (Hewitt de Alcántara, 1978; Sanderson, 1981; McGuire, 1986). As the scale of irrigation projects grew to encompass entire basins, so too did the scale and scope of social conflict (Aboites, 1998). Finally, the enormity of irrigation projects and the resultant social conflicts both revolved around and produced the need for unprecedented state intervention in the Mexican countryside, particularly after the Revolution. In short, water became a tool for rural social control, both to quell unrest and more importantly as a means of extracting surplus (Aguilar Camín, 1977; Hewitt de Alcántara, 1978; Sanderson, 1981), as well as a vehicle for rapid bureaucratic expansion (Greenberg, 1970). Its management and allocation (together with land) would also produce significant political challenges to state legitimacy, evidenced by current attempts to decentralize.

Water management in the post-Revolutionary period (1920–ca.1990) thus became increasingly federalized. However, as state intervention in the sector expanded, official resource politics produced a paradoxical effect in the Yaqui Valley. In a quite general sense, power relations became bifurcated, with the so-called social sector—*ejidos*, collectives and indigenous communities—deeply dependent upon state programmes and incorporated into official PRI political networks. Meanwhile, large landowners continued to enjoy relative autonomy *vis-à-vis* the state (Hewitt de Alcántara, 1978; Sanderson, 1981). Conflict between the two social classes[1] inevitably derived from a central paradox of Mexican history since at least the mid-19th century: how to reconcile protection of private property rights and continued private accumulation on the one hand, with agrarian reform for landless peasants and cooperative production on the other. "This challenge and the state's responses [to it]," writes Steven Sanderson, "are the keys to the developmental problems of the post revolutionary order" (1981, p. 54). Such a developmental dilemma serves as a social class dimension and counterpoint to the federal-regional tension that plays itself out in the political arena; both have directly influenced the tenor of Mexico's water reforms since 1989.

The 1934 Agrarian Code (*Código Agrario*), promulgated during the populist Cárdenas administration (1934–40), granted the federal government sweeping powers to define the 'public interest' to which water could be harnessed. According to Aboites, such authority, derived from the Revolution, gave government the ability "to intervene far more belligerently in the organization of groups involved in water use" (1998, p. 142, translation by the authors). By virtue of such legislation, between the 1930s and 1970s, the *ejido* sector and the Yaqui indigenous community were subject to direct federal control over, *inter alia*, two of the most critical components of agricultural production: water and credit (Hewitt de Alcántara, 1978; Sanderson, 1981). On the other hand, private landowners, while enjoying the benefits of federally subsidized irrigation infrastructure and guaranteed market prices, developed private credit unions and input cooperatives, and often chafed at state intervention in production (Hewitt de Alcántara, 1978).

The federal irrigation system, as it evolved in the Yaqui Valley, therefore encouraged this Janus-faced social structure, first by creating the conditions for dependency in the social sector, then withdrawing its economic and regulatory support. Over time, large landowners became highly capitalized, but by the 1970s smallholders, cooperatives and the Yaqui indigenous community were suffering from the combined effects of "capital shortage, water monopolies, underemployment, and … economies of scale" (Sanderson, 1981, p. 160). Indebted and facing private-sector control of irrigation waters, many *ejido* farmers and Yaquis were left with little choice but to rent out their land with water rights. They held accountable for unfulfilled revolutionary promises, federal bureaucracies such as the Ministry of Agriculture and Water Resources (*Secretaría de Agricultura y Recursos Hidráulicos*, SARH, previously known as SRH, which had dealt just with water but was now combined with, and assumed responsibility for, agriculture as well). Meanwhile, large landowners, with their land and water monopolies, doggedly resisted any official moves to address inequities in resource control (Sanderson, 1981; McGuire, 1986).

The Yaqui Valley case illustrates the inherently political nature of re-drawing the boundaries of water use, particularly as such re-territorialization fails to account for, or, in fact, may represent an official retreat from, historical inequities. To be fair, according to Wilder & Romero Lankao (2006), since the 1992 water reforms and decentralization, producers have noted some democratic gains in the water-management process. However, *Ejido* farmers continue to complain that they lack the financial and political muscle to check the influence of large landowners over the decision-making process. Given the historical links between official party clientelism and resource control in Mexico, the effects of the tension between local demands to decentralize decision making and official foot-dragging become more legible. The PRI's loss of official party status since 2000 has, in effect, significantly weakened the longstanding links—firmly in place since at least the Cárdenas administration—between political patronage and natural resources, particularly in the so-called 'social sector'. This is partly why Wilder & Romero Lankao have found that decentralization and privatization have allowed the state to "transfer the financial and politically charged burden of water management to non-state institutions" (2006, p. 1978). However, given the history of federal irrigation district management—the enormity and power of its bureaucracies—it is possible to understand why decentralization in water resource allocation has lagged behind that of management. Allocation, or control over the resource itself, remains a critical source of state authority in a context of declining federal power in the management realm.

In short, the conditions giving rise to contemporary efforts to decentralize water management and to the concept of river basin management were marked by extreme political and social tension at the management level, tensions that had begun to spill over into the realm of water resource control. This can be seen in the 1989 creation of the National Water Commission (*Comisión Nacional del Agua*, CONAGUA) and the formal signing of the path-breaking Lerma-Chapala Basin agreement that subsequently led to the establishment of the river basin council (*consejo de cuenca*; see Wester *et al.*, 2005). Mexico now embarked on the path of Integrated River Basin Management (IRBM). This must be contextualized with a number of additional processes that both preceded and followed the 1989 milestones.

Following from the Salinas, Zedillo and Fox years, 'integration' in Mexico's water sector has to a large extent entailed coordination of investment including planning, necessitated by serious and growing budget shortfalls for the water sector, which has increasingly been viewed as a 'hungry mouth to feed' as other more financially remunerative sectors (petroleum, manufacturing, services) have taken fiscal centre stage. Water resources management (with the proviso that planning has been implemented as an investment issue) has been neglected. River basin commissions for the Papaloapan, Grijalva, Tepalcatepec and Balsas rivers were established in the late 1940s on regional development lines, i.e. their writ covered not just water resources development, but roads, social services, etc. However, the commissions remained dependent on the federal water authority (Tortajada, 2002).

From the 1970s, a significant water management decentralization process occurred with the transfer of potable water and sewerage management to municipal governments. This was unique in Mexico's three-tier (federal-state-municipal) government structure, given that the states were largely bypassed. Water use fees that fell short of formal water rights were to be paid by municipalities to the federal government, but local water boards had such small budgets that paying direct costs for staff salaries, equipment, etc. proved difficult, a situation that continues to plague municipal water management in Mexico.

In 1992, Mexico adopted the Law of the Nation's Waters (*Ley de Aguas Nacionales*, LAN), which together with its regulations, contained specific provisions for the role of CONAGUA, the structure and functioning of river basin councils, public participation in water management, etc. Although this was a significant milestone in the modernization drive, the LAN was revised in 2003 as the underpinning for Mexico's water management regionalization initiative, which is the subject of the next section of this paper.

In 1992 there was a rapid transfer of large irrigation districts (3.2 million ha in total) from CONAGUA management to users, a process that on paper was largely complete by 1994. Given that the most critical volumes of water remained in the irrigation sector, this process was of great significance and was keenly observed from within and outside Mexico. Water users were organized at the secondary canal level and in some cases federated at the primary canal level; however, with very few exceptions, they were not permitted to formally (legally) organize at the water resource level, i.e. the reservoir or basin level. This left operation and maintenance up to the users, while the essential water allocation functions remained intact in CONAGUA's hands (Rap *et al.*, 2003). Small surface water irrigation systems and some collective groundwater systems called irrigation units (*unidades de riego*) continued under user management but posed few issues for allocation at the water resource level (see Palerm & Martínez, 2000; Silva-Ochoa, 2000; Scott & Silva-Ochoa, 2001).

It was also during the 1990s that Mexico's groundwater boom took place with rapid development and pumping of aquifers for combined agricultural, urban and industrial demand. Mexico is the largest user of groundwater in Latin America, with well over 100 000 large capacity pumps for agriculture alone. As the excesses of the boom became widely apparent, the dual competing models of groundwater technical committees (*comités técnicos de aguas subterráneas,* COTAS) and water technical councils (*consejos técnicos de agua,* also COTAS) were developed and promoted by CONAGUA and the renegade Guanajuato state water commission, respectively. Neither model has seriously taken off as a sustainable, user-driven water resources management initiative. Yet, to the credit of the seriousness of Mexican IRBM efforts, COTAS play a nascent role in the basin councils. A parallel development, invariably overlooked by water sector analysts, was the adoption of the Rural Energy Law (*Ley de Energía para el Campo*), which fixes energy

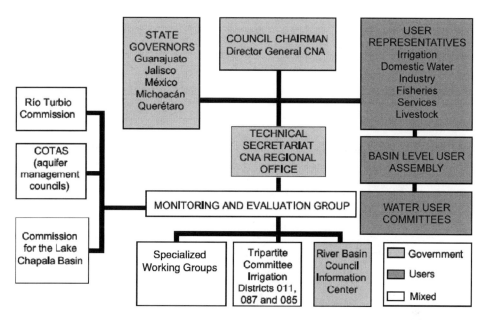

Figure 1. Structure of the Lerma-Chapala river basin

pricing for the agricultural sector and has an impact on groundwater pumping (Scott *et al.*, 2004; Scott & Shah, 2004).

It is evident that agriculture is largely in the adversarial role or position in Mexico with other (sectoral) water uses gaining ascendancy. In their most public and press presentations, urban water demands are increasingly couched as environmental imperatives, e.g. Save Lake Chapala (that is also the principal source of water supply to the city of Guadalajara). In some notable cases, stored water already allocated to farmers was released to meet urban demands (Scott *et al.*, 2001).

Since its creation in 1989, the Lerma-Chapala basin council matured significantly, formally incorporating water users from multiple sectors (Figure 1). CONAGUA has pursued similar models for basins around the country, with the initial focus on the Valle de México and the Río Bravo (Figure 2). The focus remains on river basin councils as the principal administrative vehicle to implement IWRM in Mexico (Tortajada, 2005). This is accompanied by a broader effort to decentralize CONAGUA functions through regionalization along basin lines (see below). However, commenting on the drive toward IRBM, Tortajada (2002) states:

> On the basis of their performances so far [2002], the existing basin councils cannot be considered to be viable units for water management at the regional levels. At best, they could be considered to be advisory institutions that are subordinate to the interests of CONAGUA. Fundamental changes will be necessary if they are to become successful institutions for regional water management. (p. 8)

Issues associated with the role and authority of CONAGUA with respect to water users' representatives in Mexico's IRBM process have been raised by Vargas (1999) and Mollard & Vargas (2004).

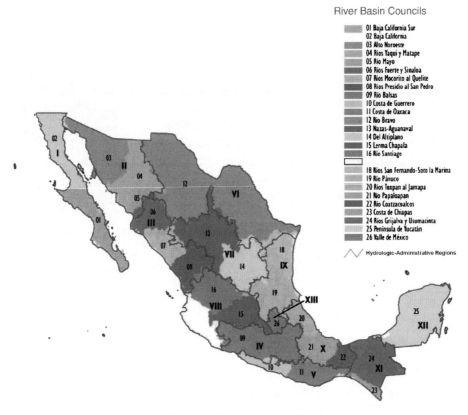

Figure 2. River basin councils

Water Management Regionalization

There does appear to be recognition in the CONAGUA for the need to devolve allocation authority over water, although this is not formally accepted. Hence the pivotal role of the CONAGUA regional office as head of the basin technical secretariat in Figure 1. How these regional offices are delineated, the multiple separate basins they encompass, and the relationship with CONAGUA federal authorities will be further explored here.

The federal-state relationship with revenue generation overwhelmingly in federal hands has hampered regional development in Mexico. Successive Mexican governments have pursued sectoral policies defined at the federal level with a weak, inarticulate or entirely absent regional focus (Tortajada, 2005). A feature of the pre-National Action Party (*Partido de Acción Nacional,* PAN) political landscape when ultimate authority was vested in the Institutional Revolution Party (*Partido de la Revolución Institucional*, PRI), this started to change through a combination of political and privatization initiatives of the PAN, and was articulated in equity and poverty-eradication terms during the 2006 abortive presidential campaign of López Obrador of the Democratic Revolution Party (*Partido de la Revolución Democrática,* PRD). The current PAN administration has yet to reconcile its underlying privatization drive with palliatives to redress social inequity. Nevertheless,

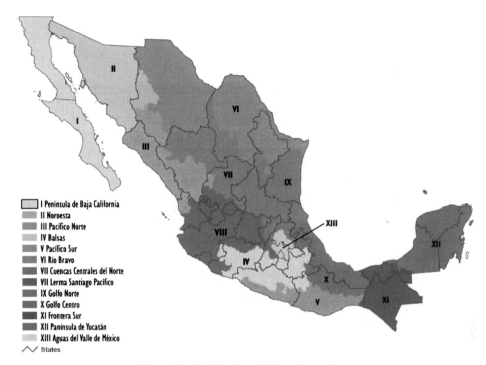

I Peninsula de Baja California
II Noroesta
III Pacífico Norte
IV Balsas
V Pacífico Sur
VI Río Bravo
VII Cuencas Centrales del Norte
VII Lerma Santiago Pacífico
IX Golfo Norte
X Golfo Centro
XI Frontera Sur
XII Paninsula de Yucatán
XIII Aguas del Valle de México
/\/ States

Figure 3. Hydrologic-administrative regions

political institutions still serve as an impediment to effective decentralization (Diaz-Cayeros *et al.*, 2002).

An important test-case for CONAGUA came with the 29 April 2004 passage of reforms to the Law of the Nation's Waters, which stipulated that the 13 decentralized CONAGUA regions (Figure 3) would become basin organizations, or more specifically, serve as the technical arm of more broad-based basin councils that incorporate civil society interests including the private sector, citizens' groups, etc. CONAGUA was given 18 months from April 2004 to publish revised regulations and establish the new basin organizations in the 13 regions. The regional delineation itself followed essentially the same lines as used by the National Water Plan Commission in 1975. These in turn closely followed an earlier 1971 hydrographic regionalization by Oscar Benassini, then SRH Director General of Studies (Melville, 2004). The 18 month period stipulated in the reform for the institutional evolution of CONAGUA regional offices as basin organizations passed without CONAGUA publishing the law's modified regulations or reorganizing its internal structure, both essential to establish river basin organizations. CONAGUA is at a critical impasse, with some officials now seeking to revise the water law yet again.

Conclusions

There are two separate and incongruous water management systems in place in Mexico. One, the official system, is derived from decades of centralized water and financial resource allocation that is firmly rooted in Mexico City. The second, a nascent form of decentralized autonomy within official institutions coupled with growing civil society demands and increasing public participation, is at loggerheads with the first. While the expeditious view is that the second will succeed from the first in an orderly process, there are clear indications that interests within official institutions, notably CONAGUA, oppose these developments.

If CONAGUA continues to retain control over resource allocation, what are we to make of the real workings of the basin councils? Are they simply a nominal approach to water resource democratization? Do they serve to create the spectre of public participation? Or, are there instances in which they have some real effect on decisions, notwithstanding their status as advisory/consensus-building forums? How might they become more of a force for change? Alternately, are they the last vestiges of entrenched regional interests, traceable—although indirectly—to the golden years of state-led development? These are critical questions for future investigation and analysis.

Certain trends are evident in political processes currently at play in Mexico. Decentralization has entailed a notch-down of arbitration and decision making from the federal to state level, where governors are carving out increasingly important power bases, just as Vicente Fox did when using the Guanajuato governorship as a springboard for his successful 2000 presidential bid. Yet IRBM with its hydrologic delineation that cuts across administrative boundaries appears to be the federal gambit to neutralize the states' growing interest and power around water resources.

Federal authority (CONAGUA) would do better to extricate itself from water resources management decision making, more effectively pursued at the local level, and facilitate river basin or 'regional' water resource allocation through data, analysis and decision-support. The allocation of scarce water (and resulting water quality deterioration) would be based on consensually agreed principles that are legally defensible based on the LAN reforms. Such a process would raise CONAGUA's technical advantage while giving voice to public opinion and a role to stakeholder involvement. However, in order to move beyond their consensus-building role, the basin councils require enhanced administrative authority currently vested in CONAGUA. This is the crossroads in the IRBM roadmap where Mexico currently stands.

Note

1. Irrigation conflicts and alliances in Southern Sonora's wheat and cotton belt also fell along the fault-lines of ethnicity, strongman leadership, official party politics, family, gender and community.

References

Aboites, A. L. (1998) *El Agua de la Nación: Una Historia Política de México (1888–1946)* (Mexico: CIESAS).
Aguilar-Camín, H. (1977) *La Frontera Nómada: Sonora y la Revolución Mexicana* (Mexico: Siglo XXI).
AWRA (American Water Resources Association) (2005) Second National Water Resources Policy Dialogue, AWRA, Tucson, 14–15 February 2005.
Biswas, A. K. (2004) Integrated Water Resources Management: a reassessment, *Water International*, 29(2), pp. 248–256.
Biswas, A. K., Varis, O. & Tortajada, C. (Eds) (2005) *Integrated Water Resources Management in South and South-East Asia* (New Delhi: Oxford University Press).
CEPAL (Comisión Económica para América Latina y el Caribe) (1999) *Tendencias Actuales de la Gestión del Agua en América Latina y el Caribe* (Santiago: CEPAL).
CONAGUA (Comisión Nacional del Agua) (2001) *Programa Nacional Hidráulico 2001–2006* (Mexico City: CONAGUA).

Diaz-Cayeros, A., González, J. A. & Rojas, F. (2002) Mexico's decentralization at a cross-roads. Working Paper No. 153 (Stanford: Stanford University Center for Research on Economic Development and Policy Reform).

Dourojeanni, A. (2003) Challenges for integrated water resources management, in: C. Tortajada, B. P. F. Braga, A. K. Biswas & L. E. García (Eds) *Water Policies and Institutions in Latin America*, pp. 13–31 (New Delhi: Oxford University Press).

Greenberg, M. H. (1970) *Bureaucracy and Development: A Mexican Case Study* (Lexington, MA: Heath Lexington Books).

Hewitt de Alcántara, C. (1978) *La Modernización de la Agricultura Mexicana, 1940–1970* (Mexico: Siglo Veintiuno Editores).

McGuire, T. R. (1986) *Politics and Ethnicity on the Río Yaqui: Potam Revisited* (Tucson: University of Arizona Press).

Melville, R. (2004) Memoria—Revista mensual de política y cultura. *Ingeniería Hidráulica en México*, 185, July.

Mollard, E. & Vargas, S. (2004) The participative management of water through basins in Mexico: lack of experience or final failure? Paper presented the Tenth Conference of the International Association for the Study of Common Property, Oaxaca, Mexico, 9–13 August.

Palerm, V. J. & Martínez, S. T. (2000) *Antología sobre Pequeño Riego Vol. II Organizaciones Autogestivas* (Montecillo, Mexico: Editores Plaza y Valdes y Colegio de Postgraduados).

Rap, E., Wester, P. & Pérez-Prado, L. N. (2003) The politics of creating commitment: irrigation reforms and the reconstitution of the hydraulic bureaucracy in Mexico, in: P. P. Mollinga & A. Bolding (Eds) *The Politics of Irrigation Reform*, pp. 57–95 (London: Ashgate Publishers).

Ruiz-Marrero, C. (2005) First People's Workshop in Defence of Water: Water Privatization in Latin America, International Relations Center, Silver City, NM, 18 October 2005. Available at http://americas.irc-online.org/am/2885

Sanderson, S. E. (1981) *Agrarian Populism and the Mexican State: The Struggle for Land in Sonora* (Berkeley: University of California Press).

Scott, C. A. & Silva-Ochoa, P. (2001) Collective action for water harvesting irrigation in the Lerma-Chapala Basin, Mexico, *Water Policy*, 3, pp. 555–572.

Scott, C. A., Silva-Ochoa, P., Florencio-Cruz, V. & Wester, P. (2001) Competition for water in the Lerma-Chapala basin, in: A. Hansen & M. van Afferden (Eds) *The Lerma-Chapala Watershed: Evaluation and Management*, pp. 291–323 (Dordrecht: Kluwer Academic/Plenum Publishers).

Scott, C. A. & Shah, T. (2004) Groundwater overdraft reduction through agricultural energy policy: insights from India and Mexico, *International Journal of Water Resources Development*, 20(2), pp. 149–164.

Scott, C. A., Shah, T., Buechler, S. J. & Silva-Ochoa, P. (2004) La fijación de precios y el suministro de energía para el manejo de la demanda de agua subterránea: enseñanzas de la agricultura Mexicana, in: C. Tortajada, V. Guerrero & R. Sandoval (Eds) *Hacia una gestión integral del agua en México: retos y alternativas*, pp. 201–228 (Mexico: Miguel Ángel Porrúa).

Scott, J. C. (1998) *Seeing Like a State: How Certain Schemes to Improve the Human Condition Have Failed* (New Haven: Yale University Press).

Shah, T., Scott, C. A. & Buechler, S. (2004) A decade of water sector reforms in Mexico: lessons for India's new water policy, *Economic and Political Weekly*, XXXIX(4), pp. 361–370.

Silva-Ochoa, P. (Ed.) (2000) *Unidades de Riego: La Otra Mitad del Sector Agrícola Bajo Riego en México IWMI*. Serie Latinoamericana No. 19 (Mexico City: International Water Management Institute).

Spicer, E. H. (1980) *The Yaquis: A Cultural History* (Tucson: University of Arizona Press).

Tortajada, C. (2002) *Institutions for Integrated River Basin Management in Latin America* (Mexico City: Third World Centre for Water Management).

Tortajada, C. (2005) Institutions for Integrated Water Resources Management in Latin America: lessons for Asia, in: A. K. Biswas, O. Varis & C. Tortajada (Eds) *Integrated Water Resources Management in South and South-East Asia*, pp. 197–318 (New Delhi: Oxford University Press).

Tortajada, C., Guerrero, V. & Sandoval, R. (Eds) (2005) *Hacia una gestión integral del agua en México: retos y alternativas, Manejo Integral* (México: Miguel Angel Porrúa).

Varady, R. G. & Meehan, K. (2006) A flood of institutions? Sustaining global water initiatives, *Water Resources Impact*, 8(6), pp. 19–22.

Vargas, V. S. (1999) *Contradicciones Socio Políticas del Proyecto de Gestión Integral de los Recursos Hídricos por Cuenca Hidrologíca: El Caso de la Cuenca del Río Laja* (Cuernavaca: Instituto Mexicano de Tecnología del Agua).

Wester, P., Scott, C. A. & Burton, M. (2005) River basin closure and institutional change in Mexico's Lerma-Chapala Basin, in: M. Svendsen (Ed.) *Irrigation and River Basin Management: Options for Governance and Institutions*, pp. 125–144, Chapter 8 (Wallingford, UK: CABI Publishing).

Whiteford, S. & Bernal, F. C. (1996) Water and the state: different views of La Transferencia, in: L. Randall (Ed.) *Reforming Mexico's Agrarian Reform*, pp. 223–234 (Armonk, NY: M.E. Sharpe).

Wilder, M. & Romero Lankao, P. (2006) Paradoxes of decentralization: water reform and social implications in Mexico, *World Development*, 34(11), pp. 1977–1995.

Small-scale Irrigation Systems in an IWRM Context: The Ayuquila-Armería Basin Commission Experience

PAULA SILVA

Introduction: IWRM Implementation in Mexico

In Mexico, there is a definitive water authority known as National Water Commission (CONAGUA)[1] who is in charge of the national water management, implying overall responsibility for all aspects of the water sector, including policy and water quality. The CONAGUA was established on 16 January 1989 as an administrative decentralized agency which aimed to induce changes in water resources utilization and management in terms of government intervention. Originally it was part of the Agriculture and Hydraulic Resources Ministry, but since 1994 it has been part of the Ministry of Environment, Natural Resources and Fisheries (in Spanish: *Secretaría de Medio Ambiente, Recursos Naturales y Pesca*, SEMARNAP).

As described in the National Hydraulic Plan 2001–2006 (CONAGUA, 2001a), the CONAGUA is divided into three levels for operational purposes: a Central Office, 13 Regional Offices in every Hydrological-Administrative Region and 20 State offices located in those states where no Regional Office is present. The Hydrological-Administrative Regions comprise the 37 Hydrological Regions in the country, which in turn cover all river basins, which number approximately 700. In 1994, a reform in the National Water Law stated that the present CONAGUA regional offices would be decentralized and transformed into autonomous basin councils (in Spanish: *Organismo de Cuenca*) and would have a Technical Council integrated by the states and municipalities in the region, which aimed to move towards Integrated Water Resource Management (IWRM).

At present, this concept is still in the implementation process. Meanwhile, it has been understood that IWRM will include all water users in the water management process and has been promoted by the CONAGUA through the formation of basin councils within every basin. Indeed, one of the objectives of the CONAGUA (2005, p. 11) is to "achieve an integrated and sustainable water management in basins and aquifers" which, among other considerations, is defined to do it through "the institutionalization of the planning process, programming, budget planning and the application of hydraulic programmes with a basin and aquifers approach". The process of social participation is mainly achieved through these basin councils which, according to the National Water Law (CONAGUA, 2004), are forums for coordination and consensus-building between representatives of the federal, state and municipal governments as well as water users of different sectors, which aim to formulate and promote the execution of programmes for improving water management in the basin.

The organizational structure within the basin councils is presented in Figure 1. The basin councils count on subordinate auxiliary organizations such as the assessment and follow-up groups, basin commissions and committees as well as aquifer management councils (in Spanish known as *Comité de Aguas Subterraneas*, COTA). The basin commissions act and convene independently but still belong to the basin council since the basin commission territory is part of the basin council. In turn, a basin committee is part of the basin commission. At present (CONAGUA, 2005) Mexico has 25 basin councils, 11 basin commissions, 16 basin committees and 69 COTAs; unfortunately, the development of these organizational structures has been halted in recent years, for example, the last basin council was formed in 2000 and the last basin commission and committees, as well as COTAS, were

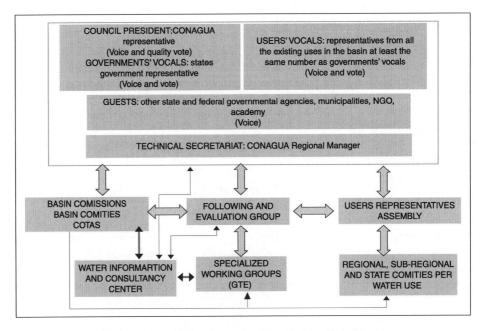

Figure 1. Organization structure of Basin Councils in Mexico

established in 2004. The formation process does not necessarily correspond to a top-to-bottom approach, i.e. first a basin council and then the required commissions, committees and COTAS. Indeed, the organizational evolution has been a response to local conditions, i.e. water competition among sectors, political purposes, social pressure, severe water crisis, etc. Even though some places lack a basin council, several commissions have been formed.

The Lerma-Chapala-Pacifico Hydrologic Region, with a total surface of 190 438 km^2, is one of the most active regions in terms of IWRM organization structure development. As shown in Figure 2, this hydrologic region is divided into three sub-regions: Lerma-Chapala, Santiago and Center Pacific. The first basin council was established in January 1993 in the Lerma-Chapala, it was also the first one in the country, and was followed by the formation of basin commissions and later by COTAS in the most over-exploited aquifer zones. The basin council in the Santiago basin was established in July 1999, but a basin council has not yet been established in the Center Pacific basin. However, two basin commissions have been created in this region: Ayuquila-Armería Rivers and Ameca River commissions. The Center Pacific basin has an area of 57 302 km^2 which is 30% of the total hydrologic region.

This paper describes the outcomes of a project related to the small-scale irrigation sector executed by the Ayuquila-Armería basin commission (AABC), which is part of this region. Albeit this commission only represents approximately 17% of the Center Pacific Region, it is a good example in order to understand the complexity of IWRM implementation and to

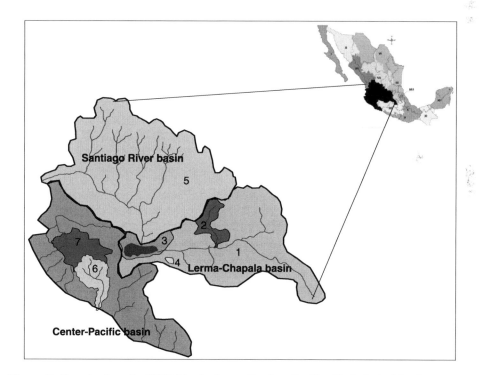

Figure 2. Organizations for IWRM in the Lerma-Santiago-Pacifico Hydrological Region. *Notes:* 1. Lerma-Chapala basin council (18 January 1993); 2. Turbio River commission (9 February 1995); 3. Lake Chapala commission (2 September 1998); 4. Parzcuaro Lake commission (18 May 2004); 5. Santiago River basin council (14 July 1997); Ayuquila-Armeria basin commission (15 October 1998); Ameca river basin commission (9 August 2004).

understand the relevance of the small-scale irrigation sector in this process. The outcomes of this experience are used to re-consider the viability of the IWRM implementation and to recognize the small-scale irrigation sector as a key element of this process.

Ayuquila-Armería Basin Commission Background

The Ayuquila-Armería basin covers approximately $9800 \, km^2$ distributed between Jalisco (82%) and Colima (18%) states. There are 30 municipalities that fall fully or partially within the area of the basin. Twenty-one are located in Jalisco state and the remainder in Colima. The basin is located between the west longitude meridians of 102° 56' and 104° 35', and the north latitude parallels 18° 40' and 20° 29'. Mexico's federal government considers it among the 15 most important rivers in the Pacific drainage area and it is one of the country's priority biodiversity conservation areas. It is important because it provides drinking water to a population of approximately 500 000 inhabitants, it irrigates over 30 000 ha of productive agricultural land and it has three major natural protected areas. A high complexity of environmental conditions characterize the watershed due to the variation in altitude and relief. Consequently, a high biodiversity exists, which includes several types of forests, one of which is the threatened cloud forest (DERN-IMECBIO *et al.*, 2003). At national level, the Ayuquila-Armería watershed is among the 43 most important watersheds with regard to biodiversity, drinking water production, irrigation surface and the presence of the Sierra de Manantlán Biosphere Reserve[2] (CONABIO, 1999).

The declaration of this biosphere reserve in 1987, which has an area of 140 000 ha (540 square miles), was mainly promoted by the Manantlan Institute of Ecology and Conservation of Biodiversity (IMECBIO) hosted by the Department of Ecology and Natural Resources of the South Coast Campus of the University of Guadalajara (DERN). This action represented the beginning of a non-governmental mobilization towards nature conservation and an integrated management of natural resources. The DERN-IMECBIO, in cooperation with the Directorship of the Sierra de Manantlan Biosphere Reserve (DRBSM), part of the National Commission for Protected Natural Areas (CONANP) of the Ministry of Environment and Natural Resources (SEMARNAT), has served as a catalyst in generating socio-political processes for improving the quality of the Ayuquila river, aiming to reduce the negative impacts of pollution on marginalized rural communities that depend on river resources, and to generate support for conservation initiatives in both urban and rural contexts (DERN-IMECBIO *et al.*, 2003).

At the same time, the CONAGUA promoted the creation of the Ayuquila-Armería Basin Commission (AABC), which was established on 18 October 1998, after a devastating environmental accident occurred when the Queseria sugar mill spilled molasses into the river. The initial goal pursued by this commission was to address the water quality problem, but due to the lack of budget the required short-term results could not be achieved directly by the AABC. However, it represented an opportunity to act as a cohesion agent for the organizations that were already working effectively in the area.

Thereafter, as an additional effort, the Inter-municipal Initiative for the Integrated Management of the Ayuquila River (IIGICRA) was established in April 2001. This organization included 12 municipal presidents, an academy organization (DERN-IMECBIO), federal government (DRSBM) and a non-governmental organization, the Manantlan Foundation for the Diversity in the Occident (MABIO). The aim of the IIGICRA was to empower local governments in the management of the natural resources and to include new mechanisms of citizen participation in decision-making processes (DERN-IMECBIO and EPFL, 2000; Martinez *et al.*, 2002). This organization formulated a proposal for environmental actions in the middle basin, which was presented to the AABC who agreed it should be included into its agenda as it represented a step forward for the success of the commission.

Finally, on 18 November 2003 a trust to provide funds to the AABC was constituted, and it expected equal contributions from the federal government—through CONAGUA—and the Jalisco and Colima state governments. However, given some administrative constraints, only state governments gave contributions and the initial funds totalled approximately US$50 000. The trust was dissolved in 2004 and the CONAGUA imbursement (an equal amount of US$50 000) was transferred to the AABC through a Collaboration Agreement, celebrated the same year by the states and the federal government. In 2005, another Collaboration Agreement was signed and all parties fulfilled their corresponding contributions with a total budget of US$100 000. Future financing is not yet strongly supported by a well-defined transfer strategy and state governments feel uncertain about CONAGUA funding. However, in the recent years the AABC has received the expected US$100 000 budget, which mainly (65%) accounts for the payment of salaries for the three full-time employees working at the AABC office and the rest is spent, according to the agenda, on training and studies (at least 12%) and the rest in equipment investment and operational costs.

The AABC Management Office—the first among all nation-wide basin councils or commissions—was established by the end of 2004 and had the responsibility to manage the budget and execute actions in accordance with the AABC agenda based on the Specialized Working Groups (GTE) action proposals as well as the commission members (see Figure 3). It was after this point that the AABC itself started accomplishing results and giving small steps towards a proper IWRM. The Irrigation Unit Inventory Update and

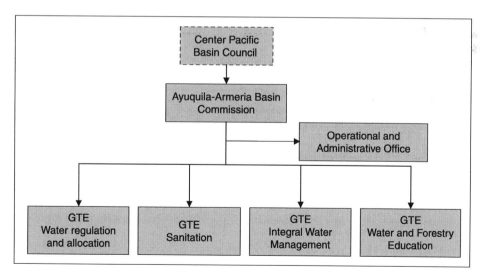

Figure 3. Organization structure of the Ayuquila-Armeria basin commission. *Note:* GTE = Specialized Working Group (in Spanish: *Grupo Tecnico Especializado*).

Diagnostic Project is one of these first steps and it was proposed by the Water Regulation and Allocation GTE and is described in the following section.

Irrigation Units: Basin Water Management Context

First, the practical implications of the theoretical concept of IWRM within the Ayuquila-Armería basin are described and, second, within this context the current situation of the small-scale system is presented in order to learn more of its role in water management and justify the implementation of the Irrigation Unit Inventory Update and Diagnostic Project.

Hydraulic Balance and Water Allocation in the Basin

As defined by the Global Water Partnership (2000, p. 22), the IWRM is:

> a process which promotes the coordinated development and management of water, land and related resources, in order to maximize the resulting economic and social welfare in an equitable manner without compromising the sustainability of vital ecosystems.

However, in general terms, the materialization of this process is at its base foundation stage, i.e. reinforcing the basin council, institutions and water legislation about which integration and optimization of the resources could be, in theory, taking place in the future. In reality, the basin IWRM perspective is mainly implemented as the water allocation among users based on the result of the study of the hydraulic balance. The CONAGUA has defined the methods to determine annual surface and groundwater availability (NOM-011-CNA-2000)[3] in hydrological regions and river basins, based on water balance principles. According to the National Water Law, the results should be used to define the granting of future water concessions within that area. The water availability of every hydraulic region is published in the official gazette (known in Spanish as *Diario Oficial de la Federacion*, DOF). Water concession titles are issued upon request if the hydraulic balance reported water availability and the process ends when the concession title is recorded in the Water Rights Public Registry (*Registro Público de Derechos de Agua*, REPDA).

In the Ayuquila-Armería basin, a study of surface water availability and hydraulic balance (CONAGUA, 2001b) was carried out in 2000 and a revision took place in 2001. Furthermore, a second revision of this study is expected by this year (2007), which will have been conducted by an external consultancy agency and supervised by the CONAGUA, but without a direct participation of the AABC or its Water Regulation and Allocation GTE. There is a special interest in the latter revision given that there is an important change in the official methodology (NOM-011-CNA-2000): the estimation of water demand will be based in the REPDA records, i.e. the actual water volume that has been properly allocated by a concession title to a user. In the previous studies, the estimation of the irrigation sector demand volume has been done by considering an average irrigation depth and the official irrigable surface.

The current hydraulic balance results showed that five sub-basins, out of the six defined for hydrological balance study purposes,[4] are over-exploited; the only unexploited sub-basin is Armería, where water drains into the Pacific Ocean. Although, at the basin level the maximum water demand does not exceed the water supply—the basin mean annual runoff is 2076 MCM and the maximum estimated water demand is 1379 MCM—there is a water shortage because the mean annual precipitation is unevenly distributed, ranging from 779 mm in Las Piedras north-east sub-basin to 1096 mm in the Armería sub-basin.

Irrigation Units

There are two types of irrigation systems in the basin, as in the rest of the country: the small-scale irrigation systems, known in Mexico as Irrigation Units (IU) and the large-scale irrigation systems or Irrigation Districts (ID). The water in the IU is supplied by means of reservoirs, derivations, pumping stations, springs, deep wells or a mixture of them; the ID are mostly irrigated by large reservoirs. The ID are directly supervised by the CONAGUA[5] and are subject to water allocation according to an irrigation plan with water measurements in control points. On the other hand, the IU are essentially independent irrigation systems managed formally or informally by the water users from their inception; governmental control of the IU only occurs when issuing their water rights and through some governmental support programmes to improve water efficiency. Before the foundation of the CONAGUA, there was a specific department for the IU oversight but it disappeared with the institutional changes.

The IU official inventory known as Irrigation Units Information System (SIUR) corresponds in most cases to field data collection from the 1970s, precisely when the IU once had some institutional control and support. This information has not been updated and discrepancies can be observed with the hydraulic balance figures and other databases related to water use records, such as the REPDA. The information reliability on IU is fragile; a comparison among available surface IU databases shows inconsistencies (see Table 1).

According to the hydraulic balance study (CONAGUA, 2001b), the irrigation sector consumes 96% of the available surface water in the basin. The IU account for 41% of the

Table 1. Data available for the surface Irrigation Units in the Ayuquila-Armería basin

Reference	Surface (ha)	Volume (MCM)	Irrigation depth (m)
Hydraulic Balance[1] (CONAGUA, 2001)	24 755	537[a]	2.17
SIUR[2]	33 567[b]	ND (728[c])	ND
REPDA[3]	ND	554[d]	ND

Notes: [1] The hydraulic balance is estimated following databases and methodology provided and approved by CONAGUA.

[2] The official inventory includes Irrigation Unit's name, type, surface and number of users (segregated by land tenure type). It has two different types of records: Organized Irrigation Units and Unorganized Irrigation Units; for the latter, the information on number of users and segregation of land tenure is not available.

[3] This means Water Rights Public Registry (in Spanish: *Registro Público de Derechos de Agua*).

[a] Represents 41% of total maximum demand volume for agriculture purposes.

[b] In accordance with the portion of the municipality within the basin and assuming that the reported irrigable area is evenly distributed along the municipality.

[c] Considering an irrigation depth of 2.17 m.

[d] This figure was estimated based on the latest REPDA records which are being used for the updated Hydraulic Balance (2007). From the water volume with agriculture concession of 1156 MCM, the Irrigation Districts water volume of 603 MCM was extracted.

irrigable surface and the ID cover the remaining irrigation sector surface of 18 278 ha in Jalisco (El Grullo module, ID 94) and 17 941 ha in Colima (Peñitas, Juarez and Tecuanillo Modules, ID 53).

The IU area reported in the last official hydraulic balance (CONAGUA, 2001b) was 24 755 ha and the estimated surface water total demand volume was 537 MCM (assuming proportional total maximum demand distribution between ID and IU), drawing an irrigation depth of 2.17 m. On the other hand, based on the SIUR municipality records, the basin IU irrigable area can be estimated to be 33 567 ha, applying the municipality area percentage within the basin to the total IU area in that municipality (assuming that its total irrigable area is evenly distributed). The surface water hydraulic balance study considered this is 8812 ha less than estimations based on SIUR. In terms of volume, this number of hectares translates into a water demand of 191 MCM (considering the irrigation depth of 2.17 m), which represents a 36% variance of the total maximum demand.

Conversely, the properly conceded water volume reported by REPDA is 554 MCM, which represents a similar volume compared to the maximum demand reported by the hydraulic balance study (CONAGUA, 2001b). This suggests that all water users have the corresponding water right, but these databases have never been cross-checked and at this point it is not known if all water users considered in the demand estimation hold a proper water right. Based on the comparison of these databases, it can be said that the IU surface water consumed in the basin is neither properly accounted for nor regulated, but there are no field data to confirm this.

Irrigation Unit Inventory Update and Diagnostic Project

Given the circumstances previously presented, the Water Regulation and allocation GTE propose to carry out an IU Inventory Update and a Diagnostic for those using surface water in each one of 30 municipalities. The IU Inventory Update and Diagnostic Project started at the end of 2004 and its implementation progress has been determined by budget availability. Until now there have been three implementation phases (see Figure 4): October to December 2004 executed in four municipalities (Tenamaxtlan and Union de Tula in Jalisco state, Armería and Comala in Colima state); from October to December 2005 executed in three municipalities (Colima, Villa de Alvarez and Coquimatla in Colima state); and during 2006, two more municipalities in Jalisco (Tuxcacuesco and El Limon).

The final project goal is to have a complete surface water IU inventory and diagnostic of the basin, but success and the timeframe depends on financial funding and CONAGUA support. In addition, the project implementation also intended to establish a first IU contact with the AABC. The activities (see Figure 5) related to methodology design, training of technicians and data analysis were conducted by members of the Water Regulation and Allocation GTE and supported by the Management Office, significantly reducing the operational and implementation project expenses. One technician per municipality was contracted to conduct fieldwork and data collection and the project was coordinated by the Management Office, supervised by members of the Water Regulation and Allocation GTE and approved by the AABC. The total cost per municipality is approximately US$2000 and expenses are mostly technicians' salaries.

A total of 93 IU were inventoried[6] (see Table 2) and a diagnostic was elaborated for each of them based on data collected at field level related to water accounting, regulation

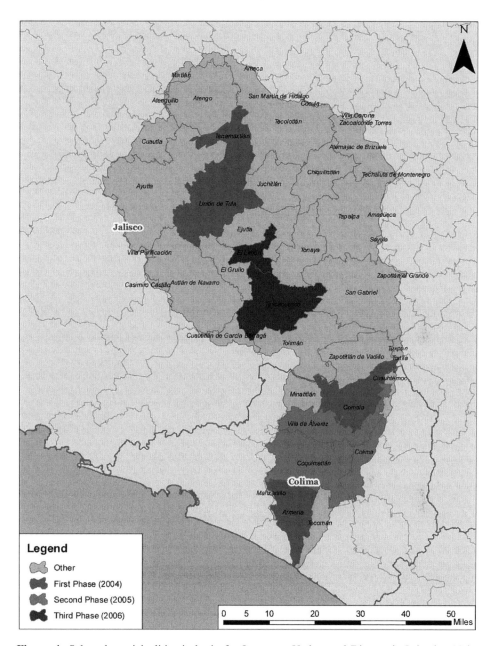

Figure 4. Selected municipalities in basin for Inventory Update and Diagnostic Irrigation Units Project

and productivity. The irrigable area covered by these IU is 12 048 ha owned by 2557 users, which represents a decrement of 14% in terms of area, and 28% in terms of number of users from the current data in SIUR. Albeit the variation on the total number of IU is not large, there are significant changes in the records since some of the IU reported by SIUR no longer exist or have changed their water use, and many others were new entries.

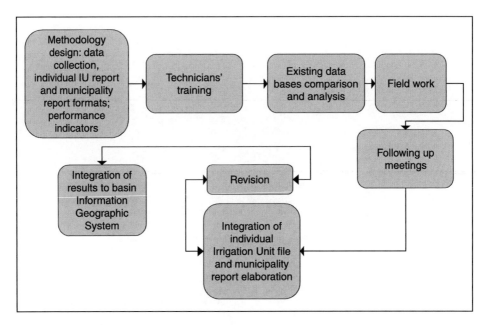

Figure 5. Irrigation Unit Inventory Update and Diagnostic Project methodology

Moreover, the total water rights in terms of volume reported by the users in the inventoried IU is 10% higher than the total volume reported by REPDA. The significance of the IU in a basin IWRM context can be appreciated based on outcomes of this project and is subsequently discussed.

Water Accounting

A water accounting procedure is required for analyzing water use patterns and trade-offs between users (Molden, 1997), i.e. water accounting is a water management tool. There are several water accounting methods currently implemented at the user level or basin level. In the small-scale irrigation systems, water accounting is empirically undertaken internally without a proper methodology or record. Therefore, information on the water volume used (frequency, duration and discharge of irrigation as well as surface irrigated) is not readily available.

In hydraulic balance studies, the calculation of maximum irrigation water demand for the IU is estimated either based on the water regulation capacity infrastructure, i.e. number of existing reservoirs, or based on the assumption that all irrigable land reported in official databases is actually irrigated and an average irrigation depth is applied. Following the latter criterion the estimated volume is 261 MCM considering 12 048 ha and an irrigation depth of 2.17 m as considered in the hydraulic balance (CONAGUA, 2001b). However, based on the data collected in the field, the volume demanded during their last irrigation season[7] in the 93 IU was estimated to be just 156 MCM, with an average irrigation depth of 1.54 m. The irrigation depth by municipality ranges from 0.6 to 2.8 m, but there are particular cases where an irrigation depth is as high as 4 m and as low as 0.1 m. Although this is a specific irrigation season estimation, it is expected to have an actual IU water

Table 2. Preliminary Inventory Update and Diagnostic Irrigation Units Project results comparison

Municipality	Number			Surface (ha)		No. of users		Water volume, water rights		Water demand estimations
	SIUR[a]	REPDA[b]	AABC[c]	SIUR	AABC	SIUR[d]	AABC	REPDA[b]	AABC[ee]	AABC[f]
Armeria	6	18	7	6618.8	5259.0	1422	1003	100.289	121.355	114.388
Comala	8	112	6	961.0	554.0	163	126	24.281	15.271	7.247
Tenamaxtlan	3	1	3	986.1	849.3	340	145	0.360	0.000	1.656
Union de Tula	4	9	5	1330.0	1375.2	315	609	10.426	10.361	9.615
Coquimatlán	38	70	26	2750.6	1904.0	703	314	42.059	31.518	13.817
Villa de Alvarez	6	37	6	194.5	218.0	2	19	6.853	2.322	0.002
Colima	20	1	1	1696.0	49.0	139	1	0.252	23.958	0.001
El Limon	12	9	7	1234.0	517.0	160	112	6.220	5.319	1.284
Tuxcacuesco	11	28	32	951.0	1322.2	193	228	11.155	11.179	7.712
Total	108	285	93	16722.0	12047.7	3437	2557	201.895	221.284	155.721

Notes: [a] Official Inventory, Sistema de Informacion de Unidades de Riego.
[b] Water Rights Public Registry, in Spanish *Registro Público de Derechos de Agua*; latest records which are being used for the updated Hydraulic Balance (2007).
[c] Ayuquila-Armeria Basin Commission records from the Inventory Update and Diagnostic Irrigation Units Project.
[d] Estimation of users for the Unorganized Irrigation Units based on the average farm plot average for the municipality.
[e] According to documentation presented by water users showing water concession title.
[f] Field estimation based on cultivated area, irrigation duration, frequency, efficiency and flows capacities according to observed hydraulic infrastructure.

demand lower than the hydraulic balance estimates since, as mentioned previously, the IU apply a smaller irrigation depth. In addition, the actual irrigated area is smaller than their irrigable land; in the best-case scenario, the IU have a cropping intensity of 1.0 and in the worst case, as low as 0.14.

Water accountability accuracy in the basin can be expected to be very low, resulting in little reliability in knowing how much water is actually being used compared to the water that has been allocated already by means of a water concession title. The question regarding this issue is to what extent an IWRM can be implemented within a basin where 40% of the volume demanded, the corresponding share of IU, has a low accuracy level of water accounting. None of the IU reported any water measurement or kept any type of irrigation record. Proper accountability is required in infrastructure and human resources, but the source of financing is difficult to determine. The IU does not have the financial resources to perform this task since most of their income is used for operational and maintenance expenses. On the other hand, the institutional tendency is to reduce governmental involvement, as occurred with the Irrigation Management Transfer programme implemented in the Irrigation Districts.

Water Regulation

In general, surface water regulation is better than groundwater regulation.[8] At present, in all municipalities where the project has been implemented, only 10 IU have been unable to provide evidence of their water right. Originally, the expected results with regard to government control over water abstractions and concession titles were more pessimistic given the first phase output that showed the overall municipalities' average for IU without a title concession was 36%. According to this first phase result, in Jalisco municipalities 50% of the surveyed IU operate without a water concession title while in Colima the situation is considerably better, with only 29%. Nevertheless, in the second and third phases of the project the situation improved significantly since 100% and 90%, respectively, of the inventoried IU had a concession title. This positive situation should not be considered to be an indication of optimal basin water regulation. Contradictory outcomes are identified when comparing REPDA records against project outputs. On the one hand, the number of records in REPDA is significantly higher than the number of IU identified during the project field work; on the other hand, according to REPDA, the total water volume with concession titles in the inventoried municipalities is approximately 10% less than the water volume the surveyed water users prove to have the rights to use.

An institutional control of water and a general overview is easier when the water users are organized and count on a formal representation in the basin commission. Unfortunately, results showed that 40% of the IU visited do not have a formal organization (i.e. *Asociación Civil, Cooperativa*, etc.) and the ones having a legal organization showed little organizational strength. In addition, the lack of organization leaves the small-scale irrigation sector without strong representation in the basin commission and with very limited access to government support programmes and private credit. It is clear that the IU are currently neglected since the level of governmental assistance is less than 19% for the case of CONAGUA and the Ministry of Rural Development (SEDER) and less than 33% for the case of SAGARPA. This situation will show little change given the current level of organization among the IU water users and the available resources in the institutions related to IU.

Water Productivity

It is not clear how the agricultural sector productivity is linked to an IWRM. Logic suggests that when productivity is low, the ability to achieve a sustainable resource management reduces. The IU water users' organizations lack statistics records for agriculture (cropping pattern, yields, production costs and agriculture production income), but farmers provided the information required for the estimation of productivity indicators.

With regard to agriculture productivity, this project reported an annual average net income of US$407/ha (US$1=11 Mexican pesos) and an average net water productivity of $1.02/m^3. There is much variation in productivity between the IU due to the cropping patterns—fruit and vegetables are sometimes as much as six times more productive than grains and pasture. In the IU of the Jalisco municipalities barley, peas and pasture are grown, except for Tuxcacuesco where important crops of water melon and other high value vegetables are produced. Meanwhile, in the Colima municipalities the cultivated crops are maize, pasture, fruits (limes, tamarind) and vegetables (papaya, tomato). Furthermore, the cultivation intensity also has an impact on productivity; in these IU on average 78% of the irrigable land is cultivated, as previously mentioned, ranging from 0.14 to 1.0, and in many cases the presence of subsistence agriculture and an absence of value-added agriculture projects is reported

Conclusions

The Irrigation Inventory Update and Diagnostic Project results show, quantitatively and with field data support, that the IUs current lack institutional control, water regulation and statistics, as well as the obsolescence of inventory data. The main findings and general conclusions are:

- The official inventory, SIUR, is completely outdated. There is an apparent reduction in the irrigable area, which could be explained by low productivity in agriculture and change of land use, i.e. from agriculture to urban use or recreational use.
- The REPDA proved to be a more precise database, but given the lack of institutional control and water accounting there is uncertainty about the actual amount of water demanded and the water users' current status, i.e. in general, concession volumes are much higher than water users are actually using, and some water users are not identified at field level. In addition, this database does not provide comprehensive information, i.e. hydraulic infrastructure, cropping patterns, etc. This information gap required to support information reliability could be provided by the IU Diagnostic conducted during this project.
- The next hydraulic balance update will probably show little variation of water availability, leading to very few opportunities for new water concession titles to be issued. Discrepancies between the water volume that is actually being accounted for in the REPDA and the water volume use in Irrigation Units will become issues if a regularization process is followed. At this point, in global terms, there is a 10% discrepancy and there will be only a 3% increase in water availability, i.e. in the current hydraulic balance (2001), 537 MCM was considered as the IU water demand and in the forthcoming version, 554 MCM will be considered.

Under these circumstances, an accurate and updated inventory becomes more relevant and information on water productivity and water management within the IU mean that IWRM has an essential role. The 93 IU diagnostics therefore represent a valuable tool for the AABC and a reference for CONAGUA to enforce the REPDA validity and to update the SIUR.

The role of IU within this basin is similar to the role nationally because across the country the percentage of demand for surface water is approximately 40%, as presented in the Ayuquila-Armería basin. The preliminary outcomes of this project can be used to re-consider the viability of the IWRM implementation. In view of the fact that the institutional overview and reliable information is absent in the IU, the basin councils are unsuccessful in meeting their goals. As a consequence, an action plan to improve water accountability, regulations and productivity conditions in the IU is essential, but is not an easy task. An analysis is required to determine sensible cost/benefits ratios of measures, or to define an acceptable percentage of unaccounted water losses because it is not possible to measure and regulate all water use in the IU.

It is suggested that the implementation of projects similar to the experience of Ayuquila-Armería should be considered as a first step towards designing a sensible and effective action plan to be implemented in the small-scale irrigation sector. The boundary conditions that have ensured its preliminary success have been:

- an end-user driven interest in water basin conservation;
- the existence of a management office;
- budget administration;
- a low-cost strategy based on institutional coordination;
- well-defined project methodology and project objectives;
- technical expertise through basin commission members (Water Regulation and Allocations GTE) since they take an active role in the project execution; and
- openness by water users to provide information (this did not always occur).

However, this project would only be considered a complete success if the following constraints could be overcome:

- certainty in the provision of financial resources directly assigned to the basin commission; and
- scrutiny, validation and approval of the project records by the CONAGUA. Without this revision, the SIUR will not be officially updated and the records cannot be used to cross-check the REPDA.

Notes

1. Its acronym was formerly CNA and has been recently changed (2005) to CONAGUA.
2. This biosphere reserve is now part of UNESCO's global network of biosphere reserves.
3. Published in the *Official Federal Daily* on 17 April 2002.
4. Officially there are only three sub-basins: Tuxcacuesco, Ayuquila and Armería.
5. The Irrigation Districts were transferred to the Water Users Associations in a national programme that started in 1989, but the headwork and the main canal (only in some cases) are still under the control of CONAGUA. The ID 094, located in the Jalisco state part of the basin, was the first large-scale system in Mexico transferred to the Water Users Associations.
6. During the field work some Irrigation Units using groundwater, i.e. deep wells, were identified, but these are not considered in the present paper

7. First phase: irrigation season 2003–04; second phase: 2004–05; third phase: 2005–06.
8. The study focused on surface water, but some groundwater IU were evaluated while in the field. Based on these findings, the lower regulation in groundwater IR could be confirmed, in most of the cases the water users were using water without a proper water right.

References

CONABIO (Comisión para la Conservación y Aprovechamiento de la Biodiversidad) (1999) *Regiones hidrológicas*. Available at http://www.conabio.gob.mx/rhp /25.html

CONAGUA (Comisión Nacional del Agua) (2001a) *Programa Nacional Hidráulico 2001–2006* (México: Comisión Nacional del Agua).

CONAGUA (Comisión Nacional del Agua) (2001b) Estudio de Disponibilidad y Balance Hidráulico Actualizado de Aguas Superficiales de la Región Hidrológica No. 12, Cuencas Cerradas de Sayula; Región Hidrológica No. 14, Río Ameca y Región Hidrológica No. 16, Armería-Coahuayana. Internal document by Altiplano de Ingeniería, S. A. de C. V. Contract Number: SGT-GRLSP 2000–2001. Mexico, D.F

CONAGUA (2004) *Ley de Aguas Nacionales y su Reglamento* (México: Comisión Nacional del Agua).

CONAGUA (2005) *Síntesis de las Estadísticas del Agua en México, 2005* (México: Comisión Nacional del Agua).

Department of Ecology and Natural Resources—The Manantlan Institute of Ecology and Conservation of Biodiversity, DERN-IMECBIO, Directorship of Sierra de Manantlan Biosphere Reserve, DRBSM/CO-NANP and Ecole Polytechnique Fédérale de Lausanne, Switzerland, LaSUR-EPFL (2003) Submitted in: ETFRN News, European Tropical Forestry Research Network, Netherlands *Global Change, Urbanization and Natural Resource Management in Western Mexico.*

DERN-IMECBIO and EPFL (2002) *Análisis integral del impacto de la urbanización sobre el Manejo de los recursos naturales. Estudio de caso: la cuenca baja del río Ayuquila, en el Occidente de México* (México/Lausanne, Switzerland: Autlán).

Global Water Partnership (2000) *Integrated Water Resources Management. TAC Background Papers No. 4* (Stockholm: GWP Secretariat).

Martínez, R. L. M., Santana, E. C. & Graf, S. M. (2002) Una visión del manejo integrado de cuencas. Curso Manejo Integrado de Ecosistemas. Colegio de Posgraduados, Montecillos, Mexico, 25 Febrero–1 Marzo.

Molden, D. (1997) *Accounting for water use and productivity*, SWIM Paper 1 (Colombo, Sri Lanka: International Irrigation Management Institute).

Integrated Water Resource Management in Colombia: Paralysis by Analysis?

JAVIER BLANCO

Introduction

There is no consensus on what the concept of Integrated Water Resource Management (IWRM) means. The Global Water Partnership (2005) considers IWRM as a means to achieve three strategic goals:

(1) efficiency to make water resources extend as far as possible;
(2) equity, in the allocation of water across different social and economic groups; and
(3) environmental sustainability, to protect the water resources base and associated eco-systems.

Traditional approaches have typically overstated one of the above goals without considering the others. For example, the ecosystem approach is directed at managing water resources, focusing on the natural ecosystems conservation and restoration (Andrade & Navarrete, 2004), but it has little regard for the allocation of water among users or the efficiency of its consumption. Engineering approaches have focused on the storage and distribution of water without considering ecosystems or land uses in the watershed.

There is no simple solution for implementing the concept of IWRM; it depends on the particular framework and institutions related to the water resources in a country. One approach that has been promoted for the implementation of IWRM in Latin America is the creation of River Basin Committees (RBC). Ideally, each RBC should be constituted by the water users in a basin, and its main function should be the allocation of water as well as the finance of activities to protect the resource. Therefore, concentrating the goals of IWRM in one institution will theoretically ensure that they will be considered equally in management of the water. Unfortunately, the creation of the river basin committees has not produced the expected results (Vargas & Mollard, 2005).

From an economic point of view, a River Basin Committee should function as a Coasian institution, fostering the direct negotiation of the users of a natural resource to achieve its optimal distribution. Coase (1960) showed that, if property rights were clearly defined and allocated to the users of a natural resource, they should achieve a social optimum through a direct negotiation. The right of water allocation is clearly assigned to a River Basin Committee, and if the water users create it, theoretically it will constitute an optimal solution for water management.

The Coasian approach has been subjected to extensive analysis with regard to the conditions where it should achieve an optimal allocation. Buchanan (1967) concluded that the direct negotiation of the users of a natural resource should be feasible only if the number of users is small (small number condition), and the transaction cost associated with administration and coordination of the negotiation should not hinder its outcome. Other barriers that might impede a Coasian negotiation are related to the dispersal or distance from the water of the users, and the lack of information about the natural resource. Therefore, it can be concluded that a River Basin Committee should function as an optimal institution for water allocation if it has the clear right of water distribution, it does not have too many water users, is closely located and has enough information about the basin.

Colombia has a unique institutional framework in Latin America, with decentralized environmental authorities and a market-based regulation for potable water and electricity utilities. In this context, the creation of RBC will generate conflicts with the functions of environmental authorities and with the regulatory bodies of water and electricity utilities.

This paper analyzes the ways of implementing the IWRM approach in the Colombian institutional context. The next section reviews the institutional framework for water management and is followed by an examination of the main instruments for water management. There is then an analysis of how those instruments can be implemented in order to achieve IWRM goals, and the final section analyzes the problems of the environmental authorities in effectively implementing them.

Institutional Framework for Water Management in Colombia

The institutional framework related to water resources in Colombia can be divided into the following categories:

(a) water allocation and pollution regulation;
(b) water demand for energy, potable water supply and sanitation, and agricultural irrigation; and
(c) ecosystem/watershed management.

The National Environmental System manages categories (a) and (c) while the water demand for sectoral use is regulated by other ministries, as described below.

The National Environmental System

Colombia has a decentralized system for environmental regulation and management. The system is directed by the Ministry of the Environment, Housing and Territorial Development, which formulates the national environmental policies, establishes the environmental regulation to be applied throughout the country, including the minimum pollution standards, and charges fees. The Ministry is also in charge of the administration of national protected areas, and grants an environmental licence to national infrastructure projects.

The regional environmental authorities (CARs) are the institutions in charge of implementing the national policies and regulations as well as managing the natural resources within their boundaries, including water resources. The main functions of CARs in relation to water resources are:

(1) to allocate water to users;
(2) to control water pollution for point and non-point sources;
(3) to formulate and adopt Watershed Ordering Plans; and
(4) to design, finance and implement activities for the protection of ecosystems.

CARs are also responsible for the conservation of forests and other ecosystems (i.e. paramos) related to the hydrological cycle.

CARs are autonomous, consisting of a regional Board not only composed of a majority of regional representatives (Department, Municipal, NGOs and Business sector, ethnic communities) but also of representatives from national government (Ministry of the Environment and the President). The Board appoints the director and approves the budget. Although CARs receive resources from the national budget, they have their own financial resources, which mainly comprise a percentage of the property tax as well as other transfers and environmental taxes generated in their jurisdiction. These characteristics give them a good degree of flexibility to allocate financial resources and implement projects according to their regional priorities.

The effective technical capacity of CARs depends mainly on the availability of resources. As the main financial resources are a percentage of the property tax, CARs that include highly populated areas have a high technical capacity and sufficient personnel.

On the other hand, CARs also have the function of reviewing and approving the environmental component of the Territorial Ordering Plans (POT). The POT is formulated by the municipalities and contains, among other aspects, the delimitation of areas for urban use/expansion and rural use. They also define categories and impose restrictions on the use of land for environmental, cultural or historic purposes.

Finally, the National Environmental System includes five research institutions:

(1) IDEAM is a national research institute which coordinates the Colombian Environmental Information System and is responsible for meteorology, hydrology, and related environmental studies.
(2) The Von Humboldt Institute is responsible for biological and biodiversity research studies.
(3) The INVEMAR Institute is responsible for marine and coastal research studies.
(4) The SINCHI Institute is responsible for research concerning natural resources of the Colombian Amazon region.
(5) The IIAP Institute is responsible for research concerning socio-economic and natural resources of the Colombian Pacific region.

Of the above, IDEAM is the main research institution related to water management. IDEAM operates the national hydrological and meteorological network, which supplies official information to all other institutions in Colombia, including aviation and hydroelectric utilities. It also establishes protocols to standardize the environmental collection of information. Among the studies undertaken by IDEAM is the National Water Study, which analyzes the water supply and demand, as well as the scarcity conditions throughout the country.

Sectorial Regulation

Water use for energy, water supply and sanitation and agricultural irrigation is regulated by sectorial ministries, as follows:

Water for domestic/municipal use. Municipal authorities should guarantee the access of potable water and sanitation to their population. Their responsibilities range from water abstraction, storage, distribution and discharge, including the treatment of residual water. Usually they grant the concession for the operation of the potable water service and sanitation to a specialized entity or utility, which could be private, semi-private or public. Each utility should have a Master Plan, which includes the main investments for the expansion of the coverage of potable water and sanitation in the urban areas. The investment and operation is financed mainly by tariffs to the users. Tariffs are subsidized for low-income users, while high-income users have to contribute above-average operating costs. In most of the municipalities, the subsidies are higher than the contributions, and therefore the deficit has to be covered by transfers from the national government to the municipalities.

The calculation of the tariffs is regulated by a national Potable Water Regulatory Committee. Tariffs should completely recover the operation costs and partly recover the investment costs of the service. The Committee has approved only the inclusion of environmental costs or investments in the tariffs limited to the charges required by law (water and pollution charges). The Ministry of the Environment designs the national policy for potable water and domestic services and helps the municipalities to reach their goals of increasing coverage in potable water and sanitation. The Ministry co-finances investments in potable water and sanitation with multilateral loans and money from public revenues. CARs also co-finance wastewater treatment systems with the pollution charge revenues.

The access or expansion of the water intake for the potable water system requires a water permit granted by the CAR with the corresponding jurisdiction. To grant a water permit, the CAR will analyze the availability of the resource and its impact to other users. The environmental law give priority for the allocation of water for domestic uses. Once the water permit is issued, the utility can access or expand the intake and will have to pay the water charge to the CAR.

Water for electricity generation. Electricity generation in Colombia works in a regulated market framework. The electric utilities are private, semi-private or public companies that compete in a 'spot' market for supplying electricity to the national grid at the lowest price. The utilities are regulated by the Energy Regulatory Committee which establishes the rules for the operation of the dispatch system. Although there is an Energy Planning Unit in the Ministry of Energy, the initiatives of new electricity plants are entirely private and not planned by the government. If a company wishes to construct an electric plant, depending on its size, it has to submit to the Ministry or to the corresponding CAR all the studies required in order to request the environmental licence. The water use permit is included in the environmental licence.

Electricity generation over 10 MW, either by hydropower or by burning fossil fuels, has to pay a 'royalty' to the CAR and to the municipality in which the reservoir or the plant is located. For the CAR the royalty equals 3% of the gross power sold, which has to be invested in financing environmental projects in the basin. This royalty constitutes the second financial source of the CARs.

Water for agriculture irrigation.   Irrigation systems for agriculture have been financed and constructed by INCODER, a specialized institute of the Ministry of Agriculture. The policy of INCODER is that the irrigation systems should be administered by their users, and the associated cost of administration and maintenance will be recovered by tariffs. Large irrigation systems have created administrative entities for their operation and maintenance. On the other hand, the lack of a responsible entity within small irrigation systems means their infrastructure has been deteriorating. In some regions, CARs and municipalities finance the maintenance of the irrigation channels.

Before initiating a new irrigation system, INCODER has to request the water use permit from the corresponding CAR.

As shown above, the water demand for domestic, energy and agricultural uses is managed through specific institutions in each sector, comprising both public and private entities. Incentives for the efficient use of the water after its intake are based on regulated tariffs, which recover from users the operating and maintenance costs of the infrastructure. Each electric company must reduce costs in order to be competitive in the spot market for supplying the national grid. The water demand for these types of uses does not take into consideration watershed planning or its effect on other users. These issues are considered by the regional environmental authorities when they approve the water uses or when applying other planning and economic instruments.

Main Instruments for IWRM

Colombia has a variety of instruments that if properly implemented could achieve Integrated Water Resource Management. Presented below is a brief description of the main instruments related to water resource management, classified in administrative, economic and planning instruments.

Administrative Instruments

Water concessions or permits.   Every person or an entity that wants to derive water from its natural environment (e.g. rivers, lakes, groundwater etc.) has to submit a request for a water concession or permit to the corresponding regional authority. The CAR evaluates the availability of the resource, the technical characteristics of the intake and the purpose of the water use, before they grant the concession. The concession is granted subject to the construction of the intake infrastructure, which guarantees that the user will not take a higher volume of water than the one authorized by the CAR.

To determine whether a volume requested by a user is efficient, CARs utilize consumption modules or factors for each type of water use (i.e. agriculture, domestic etc.). Although water concessions are granted with a 'first in, first out' criteria, if the natural resource has a critical scarcity condition, the environmental authority can declare the ordering of the resource and modify the granted concessions with a new distribution of water.

Discharges permits. Similar to the water concessions, every person or entity who wants to make a discharge into a water body has to submit a request to the regional environmental authority for a discharge permit. The environmental authority grants the permit if the discharge complies with the national standards for each pollutant. The environmental authority can only adopt a regional discharge standard if it is stricter than the national one. If the user does not comply with the national or regional standard, the authority can request the treatment of the water discharged within a compliance plan.

Environmental licences. An environmental licence is only required for projects considered to have significant impacts on natural resources. Law 99 of 1993 and Decree 1220 of 2005 and 500 of 2006 define the type of project which requires an environmental licence prior to its construction, and which of them are studied directly by the Ministry of the Environment or by the CARs. The main sectors subject to the environmental licence are hydroelectric projects, national highways, ports, mining and oil extraction among others.

Territorial Ordering Plans. Only the National Congress, the departments and municipal chambers can impose restrictions on land use. The main instrument for imposing these restrictions is the Territorial Ordering Plans (POTs). In the POTs, the municipal authorities define the category of the land (urban, suburban or rural) as well as the compatible and non-compatible uses. POTs could define non-compatible uses of rural land based on environmental reasons. If a land is severely restricted in its potential uses, the municipality has to pay compensation to the landowner. The POT has to incorporate the natural protected areas and parks declared by the national government or the regional authority. Similarly, the watershed plans constitute environmental restraints in the POTs. Finally, the environmental component of the plan has to be approved by the corresponding regional authority.

The effective enforcement of the POTs is made in granting construction licences for housing or infrastructure. For rural land uses, there is no instrument to enforce the restriction of compatible or non-compatible use of the land. For example, a plan could establish that the upper part of a watershed is non-compatible for livestock activity, but there is no mechanism or procedure for the municipality to enforce this provision.

Economic Instruments

Water use charges. Water use charges were established in Law 99 of 1993 and defined in Decree 0155 of 2004. They charge for the volume of water abstraction to all users who are granted water concessions. The water use charges are collected by the environmental authorities and CARs, and have to be invested in same the watershed in accordance with the Watershed Ordering Plan. The Ministry of the Environment defines the minimum tariff of the water use charges, and the final value is calculated by the regional authorities, depending mainly on the scarcity of the resource. The water use charge as an economic instrument should indicate what the price should be for the user if they use water efficiently.

Since the definition of the components of the water use charge in the Decree 0155 of 2004, there has been a great debate in Colombia with regard to the level of the minimum and regional tariff. On the one hand, the agriculture users argue that the level of the tariff will significantly impact the competitiveness of the sector, particularly in the context of free trade agreements with countries that have to subsidize the agriculture products. On the other hand, CARs have argued that the level of the tariff is so small that it does not cover the administrative cost of issuing and delivering the bills. In reality, both sectors are correct, taking into account that the level of the tariff is significant to agricultural users who are the main users of water, but is very small for other types of users (i.e. domestic). In watersheds with little or no irrigation, the water use charge will not generate enough income to cover the administrative expenses. The Ministry of the Environment is studying alternatives for the design of the water use charge.

Water pollution charges.   Water pollution charges were established in Law 99 of 1993 and were re-defined recently in Decrees 3100 of 2003 and 3440 of 2004. A charge has to be paid for the amount of pollution that a point source discharges in a river. The total cost to be paid will depend on a regional tariff of the charge and the volume of the pollutant discharged in a given period.

The minimal tariff is set by the Ministry of the Environment at a nationwide level and is indexed to inflation. The regional factor of the tariff depends on the overall compliance of the sources to meet a specific discharge target. The target is set at the beginning of a five-year implementation period and is evaluated every year. The target should be established taking into account the quality objectives of the water resource. Blanco & Guzmán (2005) have a complete description of the functioning of the water pollution charges in Colombia.

Planning Instruments

Watershed Ordering and Management Plans.   The actual characteristics of the Watershed Ordering and Management Plans (POMCA) were defined in Decree 1729 of 2002. Its main objective is to plan the use and sustainable management of the natural resources in a watershed in order to balance the economic benefits and the conservation of the natural ecosystems, in particular the water resources. The plan is formulated and approved by the regional authorities (CARs) following five steps: diagnosis, prospective, formulation, execution and evaluation.

The Watershed Ordering and Management Plans identify and locate the areas where it is necessary to preserve the ecosystem, change land use and/or construct infrastructure to use sustainable use the natural resources. They also include environmental goals and budgets, and define institutional responsibilities for implementing the proposed activities.

Based on the results of the POMCA, the environmental authorities can modify the granted permits (water use and pollution) as well as submit to the municipal authorities' environmental restrictions for land uses.

The Ideal IWRM in Colombia

Taking into account these instruments, it is possible construct an ideal model for implementing them guided by an Integrated Water Resource Management approach. In order to achieve the IWRM goal of the efficient allocation of water resources, CARs have to integrate three instruments: water permits, water use charges and Watershed Ordering and Management Plans. First, the Watershed Ordering Plan has to identify the water quality conditions and the type of suitable uses that it can supply. For example, the quality of water in the upper parts of the watershed are suitable for almost all types of uses, while the lower parts of the watershed could have a restriction on supplying domestic uses because of its quality. The Plan not only has to evaluate the actual condition of the water but also the desired quality conditions, based on the actual and projected land uses. As a result of this first step, the Plan has to establish clear water quality objectives as well as designated uses of water in each part of the watershed or streams. In addition, the Watershed Ordering Plan also has to establish the amount of water that can be allocated to each stream or watershed sector, taking into account the current water demand as well as a minimum ecological flow to supply biological requirements. The information about the designated uses of water with the potential supply will act as a guide for the granting of future water permits. Therefore, no water permits will be granted for uses other than the designated uses, or to volumes that exceed the potential supply.

The water use charges will function as an economic instrument to achieve the efficient allocation of the resource among the designated uses and continuously give an incentive to users to reduce their consumption of water. In order to achieve this, the tariff of water charges will incorporate the scarcity factor and charge for the effective volume that the user derives from the stream or water body. The revenues of the water use charges will finance the implementation of the activities identified in the corresponding Watershed Ordering and Management Plan.

Similarly, to control water pollution, the environmental authorities will integrate three instruments: discharge permits, water pollution charges and Watershed Ordering and Management Plans. As described above, the Plans will set the quality objectives of the water resources that are the basis for setting the pollution reduction targets for implementing the water pollution charges. Discharge permits will not be granted to an individual who affects the quality objectives of the resources.

To achieve IWRM equity goals in the allocation and quality of water resources, the process of establishing the quality objectives and the designated uses of the resource will assure the participation of all the users and communities in the watershed, and the final decision will be taken by a body with representation from the different sectors of society. Decree 1729 of 2002 established that the Watershed Ordering and Management Plan will be approved by the Board, governing body of the regional authority, and that it is integrated by representatives of municipal governments, NGOs, private sector, ethnic communities and national government agencies. The final decision of the designated uses of water as well as the water quality objectives is political in nature and has to be taken in a representative body.

Finally, the sustainability of the ecosystems associated with the water resources can be achieved by the identification of the needs for conservation or land-use changes in the POMCAs, its effective restriction in the municipal Territorial Ordering Plans, as well as restriction to project development in the environmental licensing. Not all desired land-use changes have to affect landowners and produce compensation; some activities can be achieved with cleaner production programmes, capacity building to landowners, or reforestation projects. An interesting new approach is the payment of environmental services. With this approach, the Ordering Plan can identify the ecosystems or land uses that have a clear impact on the water regulation or quality in the watershed. The regional environmental authorities could implement a payment for hydrological services to landowners who change or maintain the desired ecosystem or land use. The revenues of the water use charge and the royalties from the power generation sector could finance the payment.

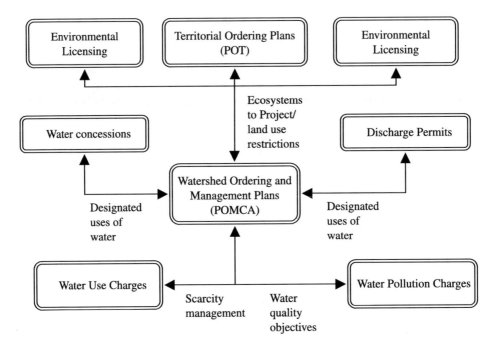

Figure 1. The relationship between instruments for IWRM in Colombia

Figure 1 illustrates how the instruments should be implemented in order to achieve an IWRM approach. Although many other types of relations may exist, Figure 1 shows the key relations between the instruments in order to achieve the IWRM goals.

Finally, information should be broadly disclosed. All users in the watershed should know key information, such as the quality objectives, the potential amount of water that can be allocated, as well as the actual supply and demand in each stream. Local information should feed into IDEAM's national environmental information system, which will then produce summary reports suitable for the Ministry to enact environmental regulation and policies.

Why Is there No IWRM in Colombia?

Despite the broad implementation of all the instruments in Colombia, it is not possible to confirm that there is integrated water resource management. The problem arises because all the instruments are implemented separately from each other and in a manner that does not generate the desired goals.

In the last three years, the environmental authorities have spent huge amounts of resources formulating POMCAs. However, a survey made by Ecoversa in 2005 of 30 regional authorities, indicates that POMCAs are mainly descriptive; they do not clearly state quality objectives for water resources or designated uses of water. They do not serve as a basis for establishing land-use restriction in the Territorial Ordering Plans or drive the target goals of the pollution charges. In conclusion, the watershed planning does not affect the implementation of the economic and administrative instruments for water management.

The implementation of both water concessions and pollution permits has the same weakness: they are issued without technical considerations of the overall impact of water abstraction or discharges into the water resource. Furthermore, once issued the environmental authorities do not verify the user compliance with the conditions set out in the permit or concession. In particular, once the water concession is granted, the environmental authorities do not verify that the infrastructure for the water intake only derives the volume authorized in the concession, and therefore, the CAR does not know how much water is actually been derived by users.

Concessions and discharge permits also have enforcement problems; only large industries comply with the standards and requirements. No small agricultural or commercial users have discharge permits. This low efficiency level is caused mainly by the low enforcement capacity of the CAR, a lack of resources to pay evaluation fees or they simply do not know the procedure.

Pollution charges and water use charges are implemented only as financial instruments for fund raising. Despite their design characteristics, the environmental authorities do not use them as economic instruments to achieve environmental goals at a minimum cost. With only the consideration of maximizing its revenues, CARs apply the water use or water pollution charges to users that do not have concessions or permits, fostering an illegal situation. Most of the municipality's discharges do not have a permit or a compliance plan but are paying the water pollution charges.

Finally, restrictions to land uses are not properly established. Law 388 of 1997 states that landowners affected by restriction to their lands have to be compensated. Therefore, the municipal authorities are not willing to impose environmental restrictions because they will impact their budgets. On the other hand, the environmental regional authorities have advanced in the declaration of protected areas, watersheds etc., but for the same reasons they do not register the restrictions to the land registries. Great confusion exists among the environmental and municipal institutions about the scope and relationship between the POMCAs and the POTs.

With regard to environmental information, usually it is confined to studies on paper and is not even available to all personnel in a CAR. The majority of the CARs do not incorporate the information in the Watershed Ordering and Management Plans in their environmental information database, and do not disclose it to the users. IDEAM does not yet have a periodic, standardized and efficient procedure to collect the relevant information from the environmental regional authorities, and the Ministry of the Environment does not give clear guidance to the research institutions for producing relevant information for decision making.

All the above indicates that although the Colombian environmental regulation has the required instruments, they are not appropriately implemented and integrated, and therefore the IWRM goals are not being achieved.

According to Castro & Castro (2002) and Ecoversa (2005), at a first glance, the cause for the erroneous implementation of the instruments is due to a lack of technical capacity by the environmental authorities. Usually CARs do not have sufficient personnel to properly implement or enforce the instruments, or they lack the expertise to undertake technical analysis (e.g. quality modelling of the resources). However, the main and hidden cause is that each instrument was designed without considering its relationship with the others, and furthermore, the Ministry of the Environment does not give clear guidance on how they should interact.

In this context, an IWRM approach can give the Ministry a coherent vision about how to formulate an integrated and harmonious regulation as well as proper guidelines to improve water resource management in Colombia.

References

Andrade, A. & Navarrete, F. (2004) Lineamientos para la aplicación del enfoque ecosistémico a la gestión integral del recurso hídrico, *Serie de Manuales de Educación y Capacitación Ambiental* (Mexico: PNUMA).

Blanco, J. & Guzmán, Z. (2006) Water pollution charges: Colombian experience, in: A. Biswas, C. Tortajada, B. Braga & D. Rodriguez (Eds) *Water Quality Management in the Americas* (Mexico City/Netherlands: Third World Center for Water Management, Springer).

Buchanan, J. M. (1967) Cooperation and conflict in public-goods interaction, *Western Economic Journal*, V, pp. 109–121.

Castro, L. F. & Castro, R. (2002) Tasas retributivas por vertimientos puntuales. evaluación nacional, Manuscript, Ministerio del Medio Ambiente, Bogotá, Colombia.

Coase, R. H. (1960) The problem of social cost, *Journal of Law and Economics*, III, pp. 1–44.

Ecoversa Corporation (2005), Ruta Crítica y Guía de Implementación de las Tasas Ambientales en Colombia— Informe Finalm, Manuscript, Ministerio de Ambiente, Vivienda y Desarrollo Territorial.

Global Water Partnership (2005), Planes de Gestión Integrada del Recurso Hídrico—Manual de Capacitación y Guía Operacional. Technical Paper (Global Water Partnership) Available at: http://www.gwptoolbox.org.

Vargas, S. & Mollard, E. (2005) *Problemas socio-ambientales y experiencias organizativas en las Cuencas de México* (México: IMTA – IRD).

The Rocky Road from Integrated Plans to Implementation: Lessons Learned from the Mekong and Senegal River Basins

OLLI VARIS, MUHAMMAD M. RAHAMAN & VIRPI STUCKI

Introduction

Integrated Water Resources Management (IWRM) has been identified as one of the basic water resources related policy approaches in several recent important commitments and recommendations, such as those of the Johannesburg Summit and World Water Forums. In the Johannesburg Plan of Implementation, the preparation of IWRM and water efficiency plans by 2005 for all major watersheds of the world was one of the two major water targets. The inclusion of IWRM as one of the standard components of Poverty Reduction Strategy Papers (the documents that are used as the baseline for targeting donor funds) has recently been discussed. This would still enhance the political role of IWRM. IWRM aims to develop democratic governance and promotes the balanced development of water resources for the reduction of poverty, social equity, economic growth and environmental sustainability.

IWRM has a history, both in agendas and in reality, but these two ends do not always meet, and sound investigations on this topic are surprisingly scarce (Lahtela, 2002; Biswas *et al.*, 2005). Different international basins have found different solutions for managing their shared waters. This paper summarizes the history and the current situation of IWRM as a theoretical concept, and contrasts this with the historical developments in the Mekong river basin in Southeast Asia and in the Senegal river basin in West Africa over the past half-century. International basin organizations are one of the key actors in both of these issues and in their development: international rivers account for 60% of all the water that flows in world's rivers. A total of 145 nations share waters with their neighbours (Wolf *et al.*, 2003).

Starting Point

Definition and History of IWRM

The philosophy of Integrated Water Resources Management has been around in various forms for several decades. The most currently used definitions include the following (GWP, 2000, p. 22, 2003, p. 1):

> ... process which promotes the co-ordinated development and management of water, land and related resources, in order to maximize the resultant economic and social welfare in an equitable manner without compromising the sustainability of vital ecosystems.

The social, environmental and economic aspects should be all be developed in an integrated and sustainable manner, under the prevalence of good governance, participation of all stakeholders and in a basin-wide context.

Over several decades, there have already been serious attempts at IWRM in different regions of the world. Many trace the roots of IWRM to the 1940s when the Tennessee Valley authority was set up to develop the water resources of that region (Barkin & King, 1986; Tortajada, 2005). However, it is obvious that the roots go far beyond that because in many countries water management has been institutionalized in an advanced and integrated way for centuries. One example is in Valencia, Spain, where multi-stakeholder, participatory water tribunals have been in operation at least since the 10th century. Embid (2003) claims that Spain was probably the first country to organize water management based on river basins because it adopted the system of *confederaciones hidrográficas* in 1926. However, it is not difficult to find other examples from literature where certain regions or countries have claimed to be the first ones in this respect. Looking back through centuries, if not millennia, it is possible to find evolved forerunners of the present IWRM paradigm (Rahaman & Varis, 2005).

The Mekong River

The Mekong river is the ninth largest river in the world if measured with runoff. With its $500\,km^3$ of water that it carries each year, it is 10 times the size of the Nile and almost double the size of all West African rivers together.

The Mekong has a catchment basin of $800\,000\,km^2$, with six countries sharing the basin (Figure 1). They are China, Myanmar, Thailand, Lao PDR, Cambodia and Vietnam. The basin has a population of almost 60 million. The GNI (Gross National Income) per capita of the riparian countries ranged between US$480 in Cambodia to US$2990 in Thailand in 2006. These figures can be compared with the GNI in Mali and Senegal, which were

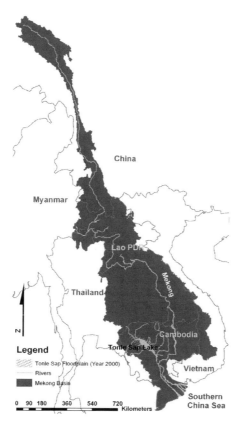

Figure 1. The Mekong river basin

US$440 and US$750 per capita (World Bank, 2007). In Vietnam, Laos PDR and Cambodia, approximately 40% of the population live below the poverty line. Over 50% of the GNI originates from fishing and agriculture.

The Mekong is one of the world's most pristine large rivers, supporting one of the most diverse and productive freshwater ecosystems (MRC, 2001).

There are major ambitions to develop the basin in various ways: by dam construction, particularly in China and Laos; agricultural development and the exploitation of forests throughout the catchment; road and settlement construction and other activities, which modify the mass flows and hydrology in a considerable way, etc.

The Senegal River

The Senegal river is a 1800 km long lifeline in the Sahel zone of West Africa. It is shared by four nations: Guinea, Mali, Mauritania and Senegal (Figure 2). The rainy uplands of Guinea are the source of a major part of the river water. It is then conveyed through the lowlands, which become increasingly arid towards the mouth of the river.

The river and the surrounding valley have supported its population fairly well through the centuries, taking into account the harsh and highly variable climatic conditions.

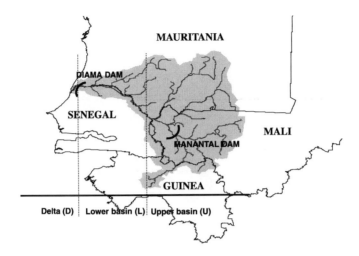

Figure 2. The Senegal river basin

This has only been possible through the use of traditional methods of livelihood and ways of using the river in cyclical matters.

However, throughout history there has been a high frequency of dry climatic periods, which have forced people to leave the valley, causing mass starvation and conflict. The last few decades have seen the augmentation of various problems in this fragile valley. Severe droughts have hit the region, the population growth rate has been extreme, the economy has declined, food security has been unstable, and, consequently, there have been numerous mass migrations, mainly to the mushrooming cities such as Dakar, Bamako, Conakry and Nouakchott.

Since the last three decades, the river has been seen as a means of enhancing the national economies of its member states. An attempt at food self-sufficiency, boosted by the problem of feeding the growing urban population and the possibility of future droughts, are the major driving forces of national and international organizations. Large-scale schemes for modernizing agriculture, hydropower generation and enabling navigation are listed as the major means of supporting such attempts. So far the success of these has been insubstantial and mostly negative. For more details and analysis see Varis & Lahtela (2002), Duvail & Hamerlynck (2003), Lahtela (2003) and Niasse *et al.* (2004).

The Early Institutions

The Mekong River Basin

The Mekong Committee was established in 1957. Its initial members were Thailand, South Vietnam, Laos and Cambodia. Burma (at present Myanmar) but China did not join. The Committee attempted to solve the regional water controversies with varying degrees of success. Its functioning was made especially difficult by China's absence, and several national and international conflicts and wars in Vietnam, Cambodia and Laos (Jacobs, 1995).

After the Second World War, which was devastating in Indochina, Cambodia, Laos and Vietnam were in an unstable situation until 1954 when they gained independence from France. The instability continued with the Vietnam War, which spread to the other countries. Cambodia has been suffering from violence until recent years.

Soon after independence, the Lower Mekong river basin countries saw the need for regional cooperation, particularly with regard to harnessing the mighty resources of the untamed Mekong. There was strong interest from the UN and US because the United Nations' Economic Committee for Asia and the Far East (ECAFE) and US Bureau of Reclamation suggested the massive development of hydropower and irrigation. In the words of the former CEO of the MRC:

> This early thinking embraced many different sectors including agriculture, navigation and flood control. Concerns of peace and security were uppermost in the minds of the planners as the world was recovering from the devastation of the 2nd World War, and former empires were breaking up into independent nations. (Kristensen, 2002, p. 2)

The Committee was the largest single attempt that the UN had made by that time. Hundreds of surveys and studies were conducted on hydrology, meteorology, topography, sedimentation, geology, fisheries, agriculture, navigation etc. to gather data and information to serve extensive development plans. A total of 180 potential large-scale projects were planned and described in this first stage of the Basin Development Plan. The sole planned hydropower generating capacity would have been 23 300 MW. Seven huge dams were planned for the mainstream Mekong and over 100 smaller dams for the tributaries. Kristensen (2002, p. 4) points out that:

> There were no organizational models to follow when the Mekong Committee began its work in 1957. Conceptually, the Committee had to deal with the challenge of formulating legal definitions of terms and concepts for water resource management entirely new to experts in international law. Given that there were no precedents to follow, the legal and conceptual achievements were substantial. One of those achievements was the 1957 Mekong Committee Founding Statute itself. No international river body had ever attempted to take on such encompassing responsibilities for financing, construction, management and maintenance of projects on an international river.
>
> In the working arrangements that were then established, National Mekong Committees coordinated activities within their country, working under the guidance and support of ECAFE and the United Nations Development Agency. An Advisory Board of international experts provided technical advice and arranged funding. The day-to-day work of the Committee was coordinated by an Executive Agent appointed by UNDP.

In its early phases, many documents refer to the Tennessee River Authority. Apparently, many experiences have been taken from that agency, given the US Bureau of Reclamation's heavy involvement in the early work of the Committee. The Mekong Committee has obviously been a model that many other basin organizations have followed in one way or another.

Despite these ambitious goals, plans and massive support, the reality took a different path. The wars came. Irrigation, navigation and hydropower plans remained as plans only. Today, malnutrition still prevails in the basin and the level of electrification has stagnated at just below 10% of the population in all areas of the catchment, besides those in Thailand. Environmental degradation and unsustainable management practices have continued to be serious and ever increasing threats to the river.

The Senegal River Basin

Guinea gained its independence from France in 1958, and the other three riparian countries of the Senegal river gained independence in 1960. The modernization of the economy of the hitherto traditional African livelihoods and economy of the Senegal valley had begun a decade earlier with the introduction of irrigated rice to the farmers. This was not successful for various reasons. It is obvious that the climatic and other natural conditions were not suitable and the economic, social and institutional structures were not suited to commercial rice farming at that time. The modern institutions started to come to the fore after independence. In the case of Senegal, the key institution since independence has been the State Development Corporation SAED (Société d'Exploitation des Terres du Delta du Fleuve Sénégal). The first international river basin organization was the OMVS (Organisation pour la Mise en Valeur du Fleuve Sénégal), which came into being in the early 1970s.

The SAED continued the colonial style policy of promoting highly mechanized irrigated rice farming on a large scale. Its focal area was on the river delta and there were very few activities upstream. The policy continued to be unsuccessful with low yields, poor market revenues, growing indebtedness, unemployment and the out-migration of large numbers of people.

The contrasts between the traditionally based small-scale irrigation-schemes and village developments (bottom-up) and the government-promoted rice irrigation schemes (top-down) became evident and started to accumulate. No comprehensive planning was undertaken; the governmental activities through SAED were centred on the commercialization of agriculture (Scudder, 1988; Adams, 1997). The scale of livelihoods, markets etc. and all the cultural and habitual factors were not properly understood and addressed in the governmental plans, which obviously ignored the bottom-up direction and thus failed.

Milestones of IWRM in 1970–89

The United Nations Conference on Water (Mar del Plata 1977)

In 1977 the UN Conference on Water was held in Mar del Plata, Argentina. Its goals were: to assess the status of water resources; to ensure that an adequate supply of quality water was available to meet the world's socio-economic needs; to increase water use efficiency; and to promote preparedness, nationally and internationally, in order to avoid a global water crisis before the end of the 20th century.

The conference approved the Mar del Plata Action Plan, which was the first internationally coordinated approach to IWRM. The plan had two parts: a set of recommendations that covered all the essential components of water management and

12 resolutions on a wide range of specific subject areas. It discussed the assessment of water use and efficiency; natural hazards, environment, health and pollution control; policy, planning and management; public information, education, training and research; and regional and international cooperation (Biswas, 2004).

The Mar del Plata conference was a success, in part due to the active participation of the developing world and the discussion concerning various aspects of water management, specifically the country and region specific analyses. The conference considered water management on a holistic and comprehensive basis, an approach recognized as one of the key IWRM issues in the 1990s. To provide potable water and sanitation facilities to all, and to accelerate political will and investment in the water sector, the conference recommended the period 1980 to 1990 as the International Water Supply and Sanitation Decade. Regrettably, transboundary water resources management was not discussed comprehensively and methods to implement the Action Plan were also ignored in the discussion process (Biswas, 2004, cited in Rahaman & Varis, 2005).

The 1980s saw the implementation of many of the Mar del Plata principles, but gradually water faded away from international agendas. For example, the Bruntland Commission Report (WCED, 1987), which laid the cornerstones of the concept of sustainable development to international policy agendas such as those of the Rio Conference in 1992, did not concentrate on water in a comprehensive way.

The Mekong River

Cambodia was absent from the Committee from the mid-1970s until 1995 due to the country's internal conflicts. During that time the Committee was known as the Interim Mekong Committee. It continued with planning activity, and published the second Basin Development Plan in 1987. Promoting hydropower was by far the primary goal.

However, once again, the plans did not materialize for various reasons, including embargos as well as internal and regional tensions in and between the basin countries. With Cambodia's absence, international projects were greatly handicapped. The massive funds required were never raised due to the instability of the region and increasingly by the growing opposition against dam construction (Kristensen, 2002).

The Senegal River

Integrated plans can be traced back to the 1970s at least in the Senegal river basin. The Senegal river development schemes date back to early 1970s when OMVS (L'Organisation pour la Mise en Valeur du Fleuve Sénégal), the river organization of Mali, Mauritania and Senegal, issued its comprehensive management plans. They included three components: irrigation, navigation and energy. *Programme integré de développement du bassin du Sénégal*, a 12-volume plan for the integrated management of the Senegal river, was first introduced in the 1974 (PNUD-OMVS, 1974). Despite this, the river valley is suffering from the consequences of serious mismanagement.

Key developments included the construction of a hydropower dam in Mali (Manantali dam) and a salt-wedge dam (Diama dam) in Senegal (e.g. Salem-Murdock & Niasse, 1996). The dams have been built, but the original three goals of the project have not been fully met (cf. Lahtela, 2002). At present the electricity and irrigation projects are still on going, whilst the navigation project is defunct.

At the same time, the SAED extended irrigation activities from the Delta to the river valley. Small irrigation schemes, PIVs (périmètres irrigués villageois), were implemented in villages. PIVs were successful during the drought years because materials and equipment were provided for the farmers free of charge. However, when support from the state ended, the schemes failed to fulfil their goals and the costs exceeded the benefits.

IWRM in 1990–99

The International Conference on Water and Environment (Dublin 1992)

Fifteen years after the Mar del Plata conference, once again water appeared on the international agenda. In January 1992 the International Conference on Water and Environment for the 21st century was held in Dublin, Ireland, to serve as the preparatory event for the Rio UNCED Conference with respect to water issues.

The Dublin conference was expected to formulate sustainable water policies and action programmes for consideration by UNCED. The conference reports set out the recommendations for action at local, national and international levels, based on four guiding principles (ICWE, 1992).

- Principle One recognized fresh water as a finite, vulnerable and essential resource and suggested that water should be managed in an integrated manner.
- Principle Two suggested a participatory approach, involving users, planners and policy makers at all levels of water development and management.
- Principle Three recognized women's central role in the provision, management and safeguarding of water.
- Principle Four suggested that water should be considered as an economic good.

The main successes of the Dublin conference were that it first focused on the necessity of water management in an integrated manner, the active participation of all stakeholders from the highest levels of government to the smallest communities and the special role of women in water management.

The major restraint of the Dublin conference was that it was mostly organized as a meeting of experts rather than as an intergovernmental meeting, and it did not consider the outcomes of Mar del Plata. Unlike Mar del Plata, there was a lack of active participation by the developing world, which was heavily criticized later. Many water professionals and decision makers of the developing world criticized the Dublin principles, especially principle four, for being vague, because without consideration of the issues of equity and poverty no water development initiatives can be sustainable (Rahaman & Varis, 2005).

The shortcomings of Dublin Principles were later addressed and tackled in the Second World Water Forum and the adjacent Ministerial Conference. In spite of several problems mentioned above, it is justified to say that current thinking about the crucial issues of Integrated Water Resources Management is heavily influenced by the Dublin Principles.

The authors consider chapter 18 of the Agenda 21, which was already prepared in December 1989 by the United Nations General Assembly and adopted at the Rio Summit, as a substantially more comprehensive document than the somewhat simplistic Dublin

Principles. Nevertheless, the Chapter 18 has been clearly in the shadow of the Dublin Principles among the international water community.

The Mekong River

By 1995, the regional political ambient became favourable to enhanced political and economic integration in Southeast Asia. As one of the results Vietnam, Lao PDR, Cambodia and Thailand signed the Mekong Agreement on the new modalities of cooperation in the Lower Mekong river basin. This agreement re-established the Committee, and it was newly named as the Mekong River Commission (MRC).

The Commission was reinforced in many ways. Cambodia is once again a member, yet China and Myanmar (formerly Burma) continue to be absent. Capacity was improved in all ways. Perhaps the major shift was from being an agency that primarily executed various projects to an agency, to one with a more strategic mandate. Instead of projects, it commits itself to a set of programmes on a more long-term basis. The vision and mission statements are:

- *Vision for the Mekong river basin:* An economically prosperous, socially just and environmentally sound Mekong river basin.
- *Vision for the Mekong River Commission:* A world class, financially secure, international river basin organization serving the Mekong countries to achieve the basin vision.
- *Mission in accordance with the 1995 Agreement:* To promote and coordinate sustainable management and development of water and related resources for the countries' mutual benefit and the people's well being by implementing strategic programmes and activities and providing scientific information and policy advice.

The programmes of the Commission are as follows:

- *Core programmes:* Water Utilization Programme (WUP); Basin Development Plan (BDP); and Environmental Programme.
- *Sector programmes:* Fisheries; agriculture, irrigation and forestry; water resources; navigation; tourism; and support programmes (focusing on capacity building of the Commission)

It is important to realize that this vision statement is in full accordance with the definition of IWRM, as quoted above. One of the major instruments, or programmes, for reaching this vision is the Basin Development Plan (BDP). It is a major exercise, scheduled over six years from 2002.

The new MRC thus continues with its activities in basin development planning. This can now be seen as the third wave of Basin Development Plans. The contemporary plans (shortened from here onwards to BDP) are different in many ways from the previous ones. This is obvious due to the more long-term, strategic type of approach that the MRC has in comparison with its predecessors.

Besides addressing different sectors such as agriculture, fisheries, hydropower etc., it has a strong focus on 'cross-cutting themes' which are: environment; human resource development; socio-economics, including poverty reduction and gender equity; and public participation. BDP should also comply with national planning practices. Cambodia has committed itself to the regional War Against Poverty, which is a broad framework for policy programmes for sustainable development and poverty reduction. BDP should comply with this framework, which guides national ministries in their activities.

The Senegal River

In 1994, the government of Senegal adopted a Master Plan for the Integrated Development of the river's left bank, which aimed to achieve the best possible compromise between social, economic and ecological imperatives. Yet, in 1995, the World Bank approved two agricultural sector related programmes presented by the Senegalese government, which did not take into account the objectives of the Master Plan accepted only one year earlier. The discrepancy becomes important when it is noted that, at the end of 1990s, SAED introduced a construction and rehabilitation programme for irrigation schemes, also contradicting the objectives of the Master Plan (Adams, 1997).

Other schemes from the 1990s include the Cayor Canal project, the Fossil Valleys Revitalization Programme and the Manantali Energy project. These are not directly aimed at IWRM but are linked to river development. At present, the Cayor Canal project has been shelved and the Fossil Valleys project is still open. The Manantali dam's hydropower station has been producing the planned 200 MW of electricity to the networks of Mali, Senegal and Mauritania since 2002 (Madamombe, 2005).

IWRM beyond 2000: More Agendas, What about Reality?

The Second World Water Forum and Ministerial Conference (The Hague 2000)

In March 2000, the Second World Water Forum was held in The Hague, the Netherlands, with more than 5700 participants from all over the world. Unlike Mar del Plata and Dublin, this Forum not only gathered together intergovernmental participants and experts, but it also saw plenty of participation from a spectrum of stakeholders related to water management from the developing and developed world. This was the key to the widely agreed satisfaction of this Forum.

With the leading theme 'From Vision to Action', the Forum brought together a vast number of vision documents structured by the World Water Council and views that were invaluable in reforming the water sector to better cope with the pressing and widely emphasized need to integrate water management. Unlike Dublin, The Hague Forum carefully considered the outcomes of previous water initiatives and acknowledged the social, environmental and cultural values of water.

The Hague forum suggested applying equity criteria and providing appropriate subsidies to the poor in the systematic adoption of the full-cost pricing of water. The Forum acknowledged that food security, ecosystem protection, the empowerment of people, risk management from water related hazards, peaceful boundary and transboundary river basin management, basic water demands, and wise water management are achievable through IWRM (Rahaman & Varis, 2005).

To meet the challenges related to IWRM, the Ministerial Declaration (WWC, 2000) called for: institutional, technological and financial innovations; collaboration and partnership at all levels; the meaningful participation of all stakeholders; establishing target and strategies; transparent water governance; and cooperation with international organizations and UN system.

'Making Water Everybody's Business' was another 'hot' topic. Privatization and public-private partnerships were widely promulgated and accepted as the means to achieve the vision objectives, which was again later opposed by many water professionals. Critics of privatization argued that the water sector is related to many functions i.e. flood control, drought alleviation, water supply, ecosystem conservation etc. and for most of these functions a government presence is necessary (Shen & Varis, 2000).

The Forum also realized that rights to land and use of water are the key determinates for people's potential to break down the poverty trap. The Forum pointed out that water could empower people, particularly women, through a participatory process of water management. Unlike Mar del Plata and Dublin, the Hague Forum discussed widely the main challenges that would implement its outcomes. After The Hague, the visions of the Forum were converted into action programmes in many of the participating countries. Global Water Partnership is now playing a central role in coordinating the 'Framework for Action'.

The Second World Water Forum was successful by putting IWRM truly in the political agenda, allowing the active participation of developing world and stakeholders, gathering world water leaders and the water community together.

The International Conference on Freshwater (Bonn 2001)

In close cooperation with the United Nations, Germany hosted the International Conference on Freshwater in Bonn from 3–7 December 2001. The aim of the conference was to contribute to the solutions of global water problems and to support preparation for the World Summit on Sustainable Development (WSSD) in Johannesburg 2002 and the Third World Water Forum in Kyoto 2003.

The Conference reviewed the previous water resources development principles and recognized that there is often a gap between making such policies and putting them into practice. The Bonn Conference tried to focus on practical ideas to find ways to implement the policies. The Conference not only identified the challenges and key targets, but also recommended actions programmes necessary to implement the policies at the field level (ICFW, 2001).

'The Bonn Keys', which summarized the conference discussions, highlighted the key steps toward sustainable development by meeting the water security needs of the poor, promoting decentralization and new partnerships. To achieve these, it suggested IWRM as the most important tool.

The Bonn Conference recommended priority actions in the fields of governance, mobilizing financial resources and capacity building and sharing knowledge. 'Bonn Recommendations for Action' addressed the issues such as poverty, gender equity, combating corruption and managing water at the lowest appropriate level. The Conference set out the actions necessary to mobilize financial resources through strengthening public funding capabilities, improving economic efficiency, and increasing official development assistance to developing countries. In the field of capacity building and sharing knowledge, it prioritized the need for education and training on water wisdom, research, effective water institution, sharing knowledge and innovative technologies. The Conference also recommended WSSD to harmonize water issues with the overall objective of sustainable development and to integrate water into the national strategies for poverty reduction (Rahaman & Varis, 2005).

The World Summit on Sustainable Development (Johannesburg 2002)

The World Summit on Sustainable Development (WSSD), held in Johannesburg, South Africa, should be recognized as a great success for putting IWRM very high on the international agenda.

The 'WSSD Plan of Implementation' includes IWRM as one of the key components for achieving sustainable development. It provides specific targets and guidelines for implementing IWRM worldwide. The major targets and guidelines include developing an IWRM and water efficiency plan by 2005 for all major river basins of the world; developing and implementing national/regional strategies, plans and programmes with regard to IWRM; improving water use efficiency; facilitating public-private partnership; developing gender sensitive policies and programmes; involving all concerned stakeholders in all types of decision making, management and implementation processes; enhancing education, and combating corruption.

It seems that the Bonn Conference recommendations that were adopted in WSSD and IWRM became by far the most internationally accepted water policy tools. The WSSD outputs encourage the major donors to commit themselves towards implementing IWRM in the developing world. A number of broad, strategic partnerships were declared along these lines. For example, the EU launched a series of strategic partnerships with regions such as Africa, Eastern Europe, Caucasus and Central Asia on Water for Sustainable Development.

The Third World Water Forum (Kyoto 2003)

Over 24 000 people from around the world attended the 3rd World Water Forum held on 16–23 March 2003 in Kyoto, Japan. The key issues discussed in the Forum were safe clean water for all, good governance, capacity building, financing, public participation and various regional issues (TWWF, 2003a). The Forum included a two-day ministerial conference, resulting in the release of a ministerial declaration on a range of water issues, including water resource management, safe drinking water and sanitation, water for food and rural development, prevention of water pollution and ecosystem conservation, as well as disaster mitigation and risk management (TWWF, 2003b).

The forum again recognized IWRM as the recommended method to achieve sustainable water resources management. The ministerial declaration addressed the necessity of equity in sharing benefits, pro-poor and gender perspectives in water policies, stakeholders participation, ensuring good water governance and transparency, capacity building of the people and institutions, developing new mechanisms of public-private partnership, promoting river basin management initiatives, cooperation between riparian countries on transboundary water issues, and encouraging scientific research.

The ministerial declaration vowed support for developing countries to achieve UN Millennium Development Goals. This declaration also commits the fullest support for developing IWRM and water efficiency plans in all river basins of the world by 2005, the target set at the World Summit on Sustainable Development. Putting various stakeholders and water ministers around the world in a Multi Stakeholder Dialogue (MSD) table together was another key feat of the Forum. A proposal to establish a new network of websites to follow up the Portfolio of Water Actions also received appreciation and support.

Different organizations and countries (i.e. World Water Council, UNESCO, UN-HABITAT, FAO, GWP, IWA, IMO, UNEP, IUCN, UNICEF, Australia, The Netherlands, EU, Japan etc.) made commitments for world water development (TWWF, 2003a).

The Fourth World Water Forum (Mexico City 2006)

The Fourth World Water Forum was held in Mexico City from 16–22 March 2006. The Forum had three guiding principles:

(1) to benefit from the value of local knowledge and experience as a key factor in the success of water policy making;
(2) to produce concrete and policy oriented outputs aimed at supporting local action on a worldwide scale; and
(3) to enable dialogue between policy sectors and stakeholders to address the complex and cross-cutting nature of water problems.

Approximately 20 000 people from around the world participated in 206 working sessions. Participants included official representatives and delegates from 140 countries, which included 120 mayors, 78 ministers, 150 legislators, 1395 journalists, experts, NGOs, companies and civil society representatives. On 22 March 2006, The Ministerial Conference adopted the Mexico Forum's Ministerial Declaration (hereafter Mexico Declaration).

The Declaration concentrated on the reinforcement of the statements of the CSD-12/13 Process by the UN Commission for sustainable Development (CSD) that came to an end slightly less than one year prior to the Mexico Forum. The focus was principally on water and sanitation issues, and other water sector concerns gained minimal attention. Rahaman & Varis (2008) critically pointed out that the Mexico Declaration failed to address the complex, interconnected and multi-dimensional facets of water management. They concluded consecutively that the Mexico Declaration is a dramatic shift in global water policy and promotes a non-holistic and fragmented water management approach.

The Mekong River

Despite the 1995 Mekong Agreement and notable international effort, the MRC has not been fully paid for by the member countries. The donor funding remains dominant, and the countries have not been investing a great deal in the MRC. The national committees remain insubstantial since ministries are unwilling to compromise their own power with the MRC. The MRC has been set aside by Thailand and Laos with regard to hydropower development, in which China is also one of the key players, since Thailand develops Mekong's hydropower resources jointly with China and Laos to help its eagerness for energy. The Asian Development Bank attracts the member countries from its partnership with the Greater Mekong Subregion (GMS) Programme, which has a large activity base on water resources. Economic integration of the region is proceeding rapidly, and the ASEAN's water programme is, again, relatively disconnected from the MRC and the GMS (Sokhem & Sunada, 2006; Mehtonen *et al.*, 2008).

The MRC's core programmes seem to be prone to certain disconnectedness, being sometimes specifically linked to particular donors (that are partly programme specific) than with one another. The BDP, as the prime agent for the integrated planning of the Lower Mekong Basin, would like to see itself above the other core programmes and sector programmes, given the content of the Mekong Agreement of 1995, but it was discontinued in 2005 due to a variety of reasons.

To follow the Johannesburg recommendations, the MRC is in the process of finalizing its Long Term IWRM Strategy (MRC, 2005).

The Senegal River

In August 2001, the West African Regional Action Plan on Integrated Water Resources Management (MEE, 2001) was presented to the possible development partners. The plan, which incorporates various international principles, those of the Dublin, Rio and Bonn conferences, for example, and of World Water Vision, includes a general framework with numerous specific objectives for IWRM in West Africa.

The reality looks rather different. The river is shared by four nations; there are two poles of development—local and national priorities, four major ethnic groups plus a number of minor ones, and several stakeholders depending on the watershed (Lahtela, 2003).

The countries still struggle with intrinsic obstacles such as the unforeseeable climate patterns and a traditionally fragmented society where ethnic groups form an important societal order. Other challenges include the underdevelopment of civil society with a low educational level, a high incidence of human poverty, adverse economic development and political instability.

The unpredictable climate, especially the threat of future droughts, continues to be a major driving force for any water use activities. However, a divergence between large and small-scale schemes is notable: the former aiming at rice production, the latter at releasing farmers from sole dependence on the rains.

The future of Senegal river development may cause major trade-offs between the national and local stakeholders. Turning these into 'win-win' solutions requires integration between the different stakeholders and institutions in society, as well as all the environmental components around the river (Niasse *et al.*, 2004). Perhaps the biggest challenge is to get the national and local levels of development to mirror their goals and actions to have the same final achievements, most importantly to get all stakeholders to express their views and affect the development options (Lahtela, 2003).

Lessons Learned

This overview of the most important milestones of IWRM at the theoretical level as articulated in major international events, as well as at the practical level in the case of the Senegal and Mekong rivers, only gives a very brief overview of this huge issue. However, it reveals the enormous mismatch and gap between the theoretical recommendations and realities of IWRM (Lahtela, 2002).

It is very obvious that the water sector should be more conscious of the present state of and future expectations of various externalities in the water sector. Too often the political, institutional, cultural, social, environmental, economic, financial and other realities turn into constraints that hamper the implementation of IWRM in practice, as the Mekong and

Senegal cases clearly demonstrate. Seeing the water issues in the broader framework of other development issues, and integrating the visions and policies of the sector, would be the way to move towards the future (Varis, 2005).

To accomplish IWRM efficiency plans for all major watersheds is an important target. It may contribute to more integrated and sustainable water management. However, plans do not help very much if proper implementation cannot be achieved. The post-Johannesburg plans should bring better results than the various previous plans. The task is anything but easy as history shows. West Africa and Southeast Asia are only two examples among many that show this. A recent study on 11 countries in South and Southeast Asia reveals that there have been many attempts and success is far from self-evident (Biswas *et al.*, 2005). The approach should be developed and elaborated much further than it is at present due to its weight in agendas as well as importance in practice.

Many of the key problems that the Mekong River Commission has faced and is facing are similar to those of the West African basin organizations. However, the agencies have attempted to solve those problems in different ways with varying success. The myriad of issues brought up by documents such as the background material for the creation of a network for West African basin organizations (GWP/WATAC, 2001; GWP/WAWP, 2002; WWF, 2002) or MRC's various strategy reports (e.g. MRC, 2000, 2001; Kristensen, 2002) can be condensed in the following 10 points (Varis, 2004; Mehtonen *et al.*, 2008; Varis *et al.*, 2006).

(1) *Strategic philosophy vs. tactical technique.* It has become clear that IWRM is a far more strategic, even philosophical issue than is often recognized. River basins are the cradles of mankind, and each basin has its own age-old and recent history. The former is a mix of cultural, ethnic, political and other factors and the latter includes institutional arrangements and governance characteristics, locally, nationally and internationally. They all influence the implementation of IWRM.

(2) *The water sector is not alone.* In the IWRM rhetoric, the water sector is typically seen as too disconnected from other sectors. The water sector itself is a multi-dimensional mosaic of activities, with no clear disciplinary boundaries (cf. Mohile, 2005). The energy, agriculture, environment, health etc. sectors are part of the water sector in the Mekong basin, but they are also sectors in their own right, and parts of other sectors. Of course, all these should be brought together, but it should also be recognized that many other sectors are suffering with similar integration challenges, with water being an important component of some of them. Seeing the water issues in the broad, cross-cutting framework of other development issues, and integrating the visions and policies of the sector, would be the way to move towards a better future through successful freshwater management.

(3) *Institutions are a grand mix.* In both the Senegal and Mekong basins, the institutional set-up is a complicated mix of various international, national, governmental and non-governmental, commercial or subsistence-related, and many other agencies and other stakeholders. The stakes and ambitions within a river basin do not originate alone from the basin itself. No single agency anywhere has the right to ignore other agencies and stakeholders. This is also the case with the MRC, ASEAN, ADB, GMS and other organizations, not to

talk about national level actors. None of the institutions has the undisputable leading role within the water sector (Sokhem & Sunada, 2006; Mehtonen *et al.*, 2008).

(4) *Stakeholder inclusion.* It is clear that all stakeholders must be included in an idealistic IWRM plan, but the stakeholders must also benefit from such participation. However, it is often very difficult to benefit all stakeholders. One of the many examples from the Mekong basin is the continuing non-membership of China in the Mekong River Commission. As shown by Makkonen (2005), China's low benefits from potentially joining the MRC are obvious.

(5) *External vs. domestic interests.* The stakes and ambitions within a river basin do not originate alone from the basin itself. The cases of the Mekong and Senegal river show clearly how diverse and massive the involvement of external powers has been in the past many decades. To a certain extent the international agencies referred to in this analysis are all driven by external actors. The donors and other actors have their own stakes, which further complicates the integrated approach to river basin management. Many of the international river basin agencies are financially dominated by donor input and do not necessarily have the full commitment and ownership of the riparian governments.

(6) *Regulation vs. development.* These two functions have too often been confused, and one agency has at the same time issued permits and regulations and been active, e.g. in dam construction, thus judging only its own operation. SAED and OMVS have not been immune to this problem. Since 1995, the MRC has developed into a strategic regulatory and planning organization with less project activities. Signs of similar developments can also be seen in West Africa. Self-regulation is ultimately an aspect that should be avoided at all costs, and societies are becoming more and more sensitive to this issue.

(7) *International river basin agencies tend to have weak coordination with other agencies, particularly with national authorities.* It appears to be common that the international basin agencies do not work easily in accord with national authorities. In West Africa, as well as in the Mekong countries, water management tasks have been delegated and split among many very fractured government departments. This seriously handicaps the implementation of IWRM. The basin agencies usually stand in a difficult position, on the one hand trying to fit their own ambitions into those of the many national governments, and on the other hand to those of the donor agencies and non-governmental organizations. Without the common recognition and ownership of the IWRM concepts in the villages, at the local governance and government levels and in the international setting, IWRM remains a theoretical concept with little sound scientific background from real-life development projects and little sustainable impact on the environment, society and economy.

(8) *Plans exist but they are not realistic and profound enough.* The MRC is infamous for its ambitious plans over decades to develop hydropower, navigation, irrigated agriculture and other economic activities in the basin with full force. However, wars, incapability of the governments, shortage of resources and recent opposition by emerging civil society organizations have stood against these plans, and they have mainly remained plans only. In West

Africa, plans are commonplace and there have been innumerable planning projects for the major watersheds. Many of them have also failed for various reasons. Obviously the plans have partly been unrealistic in terms of the institutional capacity of the nations, of their suitability to other existing plans, to their acceptability by different stakeholders and so forth.

(9) *Shortcomings in communication and participation of stakeholders.* In both West Africa and Southeast Asia, the development of open communication and public participation have had a difficult starting point. However, there is plenty of progress in most countries. The style of the MRC is evolving towards open communication, social considerations and participatory approaches are starting to be increasingly common in the investigations leading to plans. The Internet is already used in many ways and much development work is in hand. In this respect the river basin organizations have still much to learn. However, all these aspects, go very hand-in-hand with the development and respect of the civil society in these regions, and the basin agencies should recognize their high level of responsibility and make their positions clear whether they want to contribute and enhance the development of the civil society or do the opposite.

(10) *National borders cross many basins.* It is important to recognize that in the majority of developing regions of the world, IWRM requires massive international efforts because of the transboundary character of the problems, accorded typically with complicated and difficult political settings. The Johannesburg Plan of Implementation stated that all major river basins of the world should have an IWRM and water efficiency plan by the end of 2005. The Mekong and Senegal experience shows clearly that approaching the myriad of problems and challenges of the world's major river basins, many of them being transboundary, with such a one-shot plan, although being an attractive idea at the first glance, are challenging in many ways (Mehtonen *et al.*, 2008).

Acknowledgements

The authors are deeply indebted to Asit K. Biswas, Cecilia Tortajada, Pertti Vakkilainen, Marko Keskinen, Katri Mehtonen, Matti Kummu, Juha Sarkkula, Ulla Heinonen, Jussi Nikula, Mira Käkönen and the numerous Southeast Asian and West African co-workers.

References

Adams, A. (1997) Social impacts of an African dam: equity and distributional issues in the Senegal River Valley. Contributing Paper. *World Commission on Dams Thematic Reviews, Social Issues*, 1.1. Available at http://www.dams.org/docs/pdf/contrib/soc193.pdf

Barkin, D. & King, T. (1986) *Desarrollo Económico Regional (Enfoque por Cuencas Hidrológicas de México)*, 5th edición (Mexico: Siglo XXI Editores).

Biswas, A. K. (2004) From Mar del Plata to Kyoto: a review of global water policy dialogues, *Global Environmental Change (Part A)*, 14, pp. 81–88.

Biswas, A. K., Varis, O. & Tortajada, C. (Eds) (2005) *Integrated Water Resources Management in South and Southeast Asia* (New Delhi: Oxford University Press).

Duvail, S. & Hamerlynck, O. (2003) Mitigation of negative ecological and socioeconomic impacts of the Diama dam on the Senegal River wetland (Mauritania), using a model based decision support system, *Hydrology and Earth System Sciences*, 7, pp. 133–146.

Embid, A. (2003) The transfer from the Ebro basin to the Mediterranean basins as a decision of the 2001 National Hydrological Plan: the main problems posed, *International Journal of Water Resources Development*, 19, pp. 399–411.

GWP (2000) Global Water Partnership Technical Advisory Committee. Integrated Water Resources Management. TAC Background Paper No. 4 (Stockholm: Global Water Partnership Secretariat).

GWP (2003) *Integrated Water Resources Management Toolbox,* Version 2 (Stockholm: Global Water Partnership Secretariat).

GWP/WATAC (2001) *Etude de Creation de Reseau des Organismes de Bassin de l'Afrique de l'Ouest* (Ouagadougou: Global Water Partnership/Comité Technique Consultatif pour l'Afrique de l'Ouest du Partenariat Mondial d'Eau).

GWP/WAWP (2002) Rapport Introductif a l'Assemblée Générale Constitutive du REOB/AO, 10–11 July. Dakar, Global Water Partnership/West African Water Partnership.

ICWE (1992), Dublin Statement. International Conference on Water and Environment, Dublin, 29–31 December. Available at http://www.unesco.org/science/waterday2000/dublin.htm (accessed 22 June 2004).

ICFW (2001), Brief Conference Report including Ministerial Declaration, The Bonn Keys and Bonn Recommendations for Action. Available at http://www.water-2001.de/outcome/reports/Brief_report_en.pdf (accessed 20 March 2004)

Jacobs, J. W. (1995) Mekong Committee history and lessons for river basin development, *The Geographical Journal*, 161, pp. 135–148.

Kristensen, J. (2002), Incorporating international experience in river basin management: the Mekong Experience through the years. Unpublished manuscript, Mekong River Commission Secretariat, Phnom Penh.

Lahtela, V. (2002), Integrated Water Resources Management in West Africa—a framework for analysis. Licentiate Thesis, Helsinki University of Technology, Espoo.

Lahtela, V. (2003) Managing the Senegal River—national and local development dilemma, *International Journal of Water Resources Development*, 19, pp. 279–293.

Madamombe, I. (2005) Energy key to Africa's prosperity: Challenges in West Africa's quest for electricity, *Africa Renewal*, 18(4), pp. 6.

Makkonen, K. (2005) An Analysis of China's role in Integrated Water Resources management (IWRM) in South and Southeast Asia, in: A. K. Biswas, O. Varis & C. Tortajada (Eds) *Integrated Water Resources Management in South and Southeast Asia*, pp. 267–296 (New Delhi: Oxford University Press).

MEE (Ministere de l'Environnement et de l'Eau) (2001) West African Regional Action Plan for Integrated Water Resources Management. The Interim Secretariat of the Follow-up Committee for the West African Conference on Integrated Water Resources Management, Ouagadougou, Burkina Faso.

Mehtonen, K., Keskinen, M. & Varis, O. (2008) The Mekong: IWRM and cooperation between different institutions, in: O. Varis, A. K. Biswas & C. Tortajada (Eds) *Management of Transboundary Rivers and Lakes* (in press) (Berlin: Springer).

Mohile, A. D. (2005) Integration in bits and parts: a case study of India, in: A. K. Biswas, O. Varis & C. Tortajada (Eds) *Integrated Water Resources Management in South and Southeast Asia*, pp. 39–66 (New Delhi: Oxford University Press).

MRC (2000) *Strategic Plan for the Implementation of the 1995 Mekong Agreement* (Phnom Penh: Mekong River Commission Secretariat).

MRC (2001) *Strategic Plan 2001–2005 Towards and Economically Prosperous, Socially Just and Environmentally Sound Mekong River Basin* (Phnom Penh: Mekong River Commission Secretariat).

MRC (2005) *Strategic Directions for Integrated water Resources Management in the Lower Mekong River Basin, Final Draft, 12 September 2005* (Vientiane: Mekong River Commission).

Niasse, M., Iza, A., Garane, A. & Varis, O. (Eds) (2004) *Water Governance in West Africa: Legal and Institutional Aspects* (Gland and Cambridge: International Union for the Conservation of the Nature IUCN).

PNUD-OMVS (1974) *Programme intégré de développement du bassin du Sénégal* (12 volumes) (Paris: Norbert Beyrard).

Rahaman, M. M. & Varis, O. (2005) Integrated Water Resources Management: evolution, prospects and future challenges, *Sustainability: Science, Practice & Policy*, 1(1), pp. 1–8.

Rahaman, M. M. & Varis, O. (2008) Mexico World Water Forum's Ministerial Declaration 2006: A dramatic policy shift? *International Journal of Water Resources Development*, 24(1), pp. 179–198.

Salem-Murdock, M. & Niasse, M. (1996) *Water conflict in the Senegal River Valley: Implications of a 'no-flood' scenario*. Paper No. 61 (Binghampton, NY: International Institute for Environment and Development).

Scudder, T. (1988) *The African Experience with River Basin Development: Achievements to Date, the Role of Institutions, and Strategies for the Future* (Binghampton NY: Institute for Development Anthropology).

Shen, D. & Varis, O. (2000) World Water Vision: balancing thoughts after The Hague, *Ambio*, 29, pp. 523–525.

Sokhem, P. & Sunada, K. (2006) The governance of the Tonle Sap Lake, Cambodia: Integration of local, national and international levels, *International Journal of Water Resources Development*, 22, pp. 399–416.

Tortajada, C. (2005) Institutions for IWRM in Latin America, in: A. K. Biswas, O. Varis & C. Tortajada (Eds) *Integrated Water Resources Management in South and Southeast Asia*, pp. 297–318 (New Delhi: Oxford University Press).

TWWF (2003a) Summary Forum Statement. The Third World Water Forum, 16–23 March. Available at http://www.world.water-forum3.com/en/statement.html (accessed 10 October 2003).

TWWF (2003b) Ministerial Declaration. The Third World Water Forum-Ministerial Conference, 16–23 March. Available at http://www.world.water-forum3.com/jp/mc/md_final.pdf (accessed 25 September 2003).

Varis, O. (2004) Basin organization models: the case of the Mekong River, in: M. Niasse, A. Iza, A. Garane & O. Varis (Eds) *Water Governance in West Africa: Legal and Institutional Aspects*, pp. 223–238 (Gland and Cambridge: International Union for the Conservation of the Nature IUCN).

Varis, O. (2005) Externalities of integrated water resources management in South and Southeast Asia, in: A. K. Biswas, O. Varis & C. Tortajada (Eds) *Integrated Water Resources Management in South and Southeast Asia*, pp. 1–38 (New Delhi: Oxford University Press).

Varis, O. & Lahtela, V. (2002) Integrated water resources management dilemma along the Senegal River— introducing an analytical framework, *International Journal of Water Resources Development*, 18, pp. 501–521.

Varis, O., Kummu, M., Keskinen, M., Makkonen, K., Sarkkula, J. & Koponen, J. (2006) Integrated Water Resources Management on the Tonle Sap Lake, Cambodia, *Water Science and Technology: Water Supply*, 6(5), pp. 51–58.

WCED (1987) *Our Common Future: Report of the World Commission on Environment and Development* (Oxford: Oxford University Press).

Wolf, A. T., Yoffe, S. B. & Giordano, M. (2003) International waters: identifying basins at risk, *Water Policy*, 5(1), pp. 29–60.

World Bank (2007) *World Development Indicators* (Washington DC: The World Bank).

WSSD (2002) *Report of the World Summit on Sustainable Development* (A/Conf. 199/20). Available at http://www.johannesburgsummit.org (accessed 14 December 2003).

WWC (2000) *Final Report. Second World Water Forum and Ministerial Conference. Vision to Action* (Marseilles: World Water Council) Available at www.worldwaterforum.net

WWF (2002) *African Rivers Initiative: Concept Paper* (Gland: World Wildlife Fund).

Capacity Building: A Possible Approach to Improved Water Resources Management

ALEXANDRA PRES

Introduction

In the aftermath of the Rio Earth Summit, the term 'capacity building' was born, reflecting the great need for performance improvement at individual, organizational and sectoral levels to contribute to sustainable development. This is not the only dimension of reflection: capacity building is a mirror for a wide range of trends and ideas within a highly complicated and sophisticated system; a mirror of the system itself, its achievements, but also its failures and contradictions. Therefore, the nature of capacity building is very complex. As a consequence, this paper cannot cover its whole and continuously emerging complexity, but it will outline some definitions and trends, the current challenges and a possible approach of capacity building to contribute to improved water resources management.

This will be done based on the following perceptions:

- Having invented the term 'capacity building', the international community has agreed that people instead of plans or structures are drivers of change and performance.
- Agreeing on the importance of capacity building to ensure development and a sustainable environment does not mean that capacity building is a panacea that makes it possible to easily overcome bottlenecks and failures of the past.

Trends and Definitions

Since early 1950s, the need to enhance capacities had been recognized in the context of development aid. However, capacity building was reduced to training only.

Most frequently, new infrastructure, technology or management concepts were the starting point for the very first type of human resources development. Thus, the focus was more on what type of skill X employee or technician Y required to be able to run plant Z in an appropriate manner. Processes were almost neglected and interventions were done punctually, widely spread and on an ad hoc basis. As a consequence, training was mostly of a specialized and diversified nature, providing individuals with technical tools and equipment to improve their performance.

During the 1970s, the role of institutions was recognized and they became the focus of capacity building interventions; organizations were restructured, re-designed and strengthened. Training became one tool for human resource development that aimed to create a capable citizen (Kuehl, 2004). This trend highlights an important change within the capacity building perception; strengthened, trained individuals could not fill development gaps per se. Thus, interventions were linked to their organizational surrounding.

The trend continued in the 1980s, when the processes and links between organizations were involved in capacity building measures that became more and more holistic and process-oriented. Finally, the importance of the inter-linkage of the individual–organizational–system levels has now become state-of-the-art (Kuehl, 2004). Therefore, the importance of institutional changes and the impacts at the system level have increased and they are currently the priority area of interest.

It is obvious that the complexity of capacity building has increased enormously during the last few decades, which has also brought a variety of definitions and semantic interventions. Some even judge the term 'capacity building' as no longer politically correct because its connotation suggests that there is no capacity at all and thus contradicts new paradigms of development aid. Thus capacity development or enhancement is mentioned in line with capacity building. So what are we talking about?

According to OECD-DAC (2006), 'capacity' means the ability of people, organizations and society as a whole to successfully manage their affairs, and 'capacity building' means a process whereby people organizations and society as a whole unleash, strengthen, create, adapt and maintain capacity over time. Certainly, this is a most abstract definition, which is agreed by most players.

Internationally, capacity building (or capacity development) is now considered to be one of the critical factors missing in current efforts to meet the Millennium Development Goals, and at the same time is considered an important challenge fraught with difficulties in meeting it. OECD/DAC considers that, until recently, capacity building was viewed as a technical process of transferring knowledge or organizational models from North to South, and has been insufficient in considering the importance of country ownership. The Paris Declaration of 2005 now postulates that capacity building is an endogenous process, with donors playing a supporting role, and highlights the role of local ownership in the context of development.

Therefore, this definition reveals that there is no standard format, but rather a search for a best-fit solution under particular circumstances. Furthermore, it reveals the fact that capacity building is method, result, aim and mean at the same time and it particularly reveals one dilemma: local ownership is required to build capacity, but capacities are required to build ownership. So how is InWEnt, as a capacity building institution of the German Development Cooperation, defining capacity building?

InWEnt's definition is simple. Within an international context capacity building encompasses advanced professional training, dialogue, networking and advisory services for human resources development with the aim of strengthening the capacities of partners to plan and implement sustainable development strategies and policies (InWEnt, 2006a,b). Thus, it is obvious that InWEnt covers only one part of capacity building, and can be considered as a complementary instrument to financial and technical assistance that definitely contributes to capacity building, too. To go further, it is a successful development cooperation that requires the complementary intervention of three instruments: financial, technical and personnel assistance.

Apart from being a challenge by itself, what are the biggest challenges that capacity building faces?

Challenges

In general, the major challenge can be summarized in two words: continuous change. It might sound simple, but the dimension of change is very complex and manifold:

- The world has become unpredictable and insecure: unanticipated national disasters and political conflicts appear more frequently, which have severe effects on development.
- The world has become faster: new developments, e.g. in the field of information technology, allow an enormous and rapid dissemination of information and discoveries that have to be considered in future actions.
- The world has become unbalanced: the social gap between developed regions and developing regions is increasing.
- The world is becoming 'smaller': 'local' and 'global' are not contradictions any more, globalization has an impact at a local level, but on the other hand, a local crisis can affect the global level.
- The world has become cooperative: the global economy and environment request intercultural cooperation and the understanding of different values and mindsets.

This list exemplifies the complexity of change and reveals an utmost challenge that is faced by individuals, organizations and systems, which have to act under conditions of continuous change, the only persistent paradigm mankind can refer to.

Based on this general assumption that refers to any human system, therefore to any organization and sector, water sectors and their organizations currently face the following main challenges.

Water sector organizations are often highly professional bodies, following highly professional values and are doing a proficient job in their respective technical fields. However, competence is at the core of many aspects of management, which is bringing with it increasingly numerous managerial, inner- and inter-institutional challenges. The organizations face multi-faceted and multi-level institutional arrangements. They operate at national, regional and local levels. They are influenced by neighbouring countries and regions, as well as by international actors. They have functional differences and links, e.g. between regulatory bodies and drinking water providers. They are influenced by the interests of sectors such as agriculture, industries or tourism, to name but a few. They are equally influenced by political interests at all levels of administration and have to maintain respective links to all of them.

Furthermore, a great variety of challenges have to be addressed in a complementary manner, e.g. good governance with regard to resource management, effects of urbanization and simultaneous ruralization tendencies, the increasing pollution and decreasing quality of the resource, the requirements of food security and health, crises prevention and management, changes in demographic patterns and, last but not least, effects of globalization.

Being embedded in such a complex system, even a variety of systems, implies that organizations in the water sector need to respond to an enormously complex chain of causalities and a multi-level, holistic environment. Therefore, water management or water sector management is far from being limited to technical manoeuvres.

It is not only the environment that is highly dynamic, but also the water sector organizations, and they are currently undergoing massive changes; the institutional landscape is under an enormous restructuring processes. Decentralization assigns new tasks and roles to old organizations and comes together with new subsidiary entities. Organizational structures and practices change in response to new paradigms such as Integrated Water Resource Management. Privatization brings about new roles together with ethical questions. In this change, organizations carry the burden of their past. They tend to remain:

- bureaucratic and hierarchically organized (over-regulated);
- over-staffed and underpaid, resulting in a lack of drive and motivation;
- technically-oriented instead of future- and managerial-oriented; and
- governed by poor decision making and contradicting priorities.

The capacities to deal with the complex and changing sector and its environment are poorly developed. Organizations in the water sector are frequently headed by engineers or former technicians, whose entry into managerial positions was through technical performance. Civil service personnel are not experienced nor trained for entrepreneurship in autonomous or semi-autonomous bodies and do not easily follow the shift towards autonomy and a client orientation. As a consequence, in most cases, the modus operandi continues to be technically-oriented, neglecting the fact that water is no longer a resource-based problem but one that is management-based.

How to can the described problems be overcome? How can the water sector act under continuous change?

A Possible Approach

A modern capacity building programme has to follow a systemic approach and has to provide a set of multi-level interventions by applying different instruments, which have to be tailor-made to the target group(s) of interest. In order to be able to transfer knowledge and skills, content has to be combined with methods that allow its application. Consequences of change are changing needs. Therefore, capacity building programmes have to be designed in a flexible manner. A flexible approach allows the focus and activities of a programme to be adapted according to changing needs and any feedback given by all parties involved, in particular by the partners on site. Furthermore, it allows the extension of activities in combining and harmonizing activities with those of other capacity building actors and thus, as a consequence, might increase impacts by joining efforts.

In practice, InWEnt capacity building programmes resemble a set of bricks that might consist of the following six components that are closely linked to each other:

- Component 1: Professional knowledge
- Component 2: Methodical competence

- Component 3: Regional cooperation
- Component 4: (Training) needs assessment, monitoring + evaluation
- Component 5: Public relations and public awareness
- Component 6: Community of practices

Figure 1 shows the above-mentioned programme design developed and applied within the water portfolio of InWEnt.

The idea behind these bricks is simple. Component 1 focuses on a knowledge up-grade highlighting managerial aspects of water resources management (e.g. business + strategic planning, financial management). Component 2 provides a wide range of methodical competence (e.g. training of trainers, moderation skills, coaching aspects). The core activity of this component is the management of development and change processes, particularly highlighting those processes in the context of human resources development and organizational performance improvement.

Component 3 aims to strengthen cooperation and networking between water sectors in a particular region. It is based on the conviction that problems and bottlenecks are often of a similar nature and that there are enough solutions on site, technical and managerial, but the knowledge is lacking about their existence. The comparative learning allows partners to share knowledge and experience that can be the basis for regional solutions and approaches, e.g. in the context of water governance. Thus, this component aims to contribute to an impact at the system level. It also serves as one feedback loop for the detailed activity planning of Components 1 and 2. By learning from partners, it contributes to local ownership because the feedback given can be reflected in the future planning and implementation of the programme.

Component 4 allows a third-party view and can be considered as a further feedback loop to the programme. Experts and scientists continuously assess newly-appearing needs and monitor the process of implementation. Component 5 focuses on the population in a broader sense and aims to create awareness of the preciousness of water resources and their efficient use. The method of dissemination is completed by Component 6, which aims to share good practices, as well as bottlenecks and challenges faced during the programme's implementation, with the international water community.

A programme design cannot improve performance by itself, but sets a sound basis for doing so. Instruments applied, contents delivered and organizations selected are the most

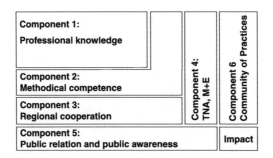

Figure 1. InWEnt's capacity-building approach in practice.

important components of a capacity building programme. InWEnt is focusing on the key players within the water sectors covered. Instead of tackling them all, a few were carefully selected and the principle is followed that less is more. Contents vary, but always focus on the strengthening of management capacities towards an efficient and sustainable use of water resources. With regard to instruments, dialogue and networking have become more important during the last few years. It does not mean that advanced training is no longer required, but that the need for dialogue between various stakeholders plays a major role to reduce conflicts of interest. The same applies to networking that allows a cost-efficient method of learning by exchanging existing knowledge, the asset of comparative knowledge so to speak.

What are the other lessons learned that are worth sharing?

Lessons Learned

A flexible approach has certain advantages. It continuously allows the appropriate adaptation and close involvement of all parties concerned. Furthermore, the extension of intervention can easily be carried out. Such types of extension might refer to the involvement of a further target-group and/or a further target-country, but might also refer to the strengthening of particular components of funded activities by third parties. On the other hand, a systemic approach that combines professional knowledge and methodical competences is a prerequisite to strengthen the capacity to act and decide, a skill that is highly necessary to be able to act within a continuously changing environment.

Furthermore, it can be seen that there is a renewed interest in strengthening inter-organizational networks. In this context, regional approaches of capacity building programmes gain a particular value.

Such complex programmes are time-consuming due to a high need for coordination between all actors involved. Thus, its complex nature requires sound project management and efficient in-house procedures, as well as a rapid in-house feedback mechanism to continuously improve implementation. In particular this refers to the coordination or feedback mechanism between sectoral and methodical knowledge. Many acknowledge that there is a need for an improvement in managerial performance by the water sectors, its organizations and policy makers, but a few also consider that the need for change in management is necessary and unavoidable. Change is still a most sensitive issue, which the further development of capacity building approaches has to focus on. In this context, strengthening of global governance competence will increasingly play an important role. As a consequence, greater emphasis will have to be given to strengthening driving forces and leaders. This might be the future challenge, but a chance for capacity building, too.

One aspect is most evident: there is an enormous need for capacity building at all levels. Therefore, it is definitely time to join the efforts of different players in the field and it is also time for new approaches. If the current water programme of InWEnt still focuses on advanced training, dialogue and networking, future programmes have to include the strengthening of existing training entities on site. By doing so, these entities might cover a wide part of training needs and institutions, just as InWEnt might focus on process moderation, leadership training and network facilitation.

Summary and Conclusion

Capacity building has evolved over many years, growing with the complexity of society and systems. Being a learning concept, so far it has been protected from becoming a further paradigm. However, it faces enormous challenges: the strengthening of people that have to adapt to, make decisions and act under continuous change. Being most in vogue, it faces the pressure and abuse of being made a scapegoat whenever static, bureaucratic and 'one-size-fits-all' policies cannot be implemented.

Capacity building has many challenges: it is a concept with high potential as it puts people at the core of its interest. People are not static by nature; people are able to perform, which as consequence means dynamism. Due to the nature of dynamism the 'what' is difficult to anticipate, even if it seems clear to most of us; but the 'who' can be defined. Thus, it might be worth taking time to look more closely at some other disciplines (Varis *et al.*, 2008) that describe the fundamental structure of systems and people and their behaviour to learn more about the interlinking 'how'.

As long as this 'how' remains a mystery, people will focus on the daily business and get caught up with complexity at an operational level; they will rather fulfil tasks than perform. Improved water resources management might be achieved by fulfilling tasks. An integrated water resources management requires more: it requires social commitment and future-oriented interdisciplinary performance; a conviction that is based on human welfare and thus the change of individual attitudes towards a common one, and it requires a committed leadership that is inclusive rather than exclusive.

Capacity building might not have the answers yet, but it is able to ask some of the right questions.

Acknowledgements

The author is deeply indebted to Asit K. Biswas for being a mentor and for being so patient. Sincere respect goes to Hinrich Mercker and Hans Pfeifer for many reasons, and special thanks go to the most sincere colleagues of the partner organizations the author has had the pleasure of working with.

References

InWEnt (2006a) *Capacity Building Concept* (Bonn: InWEnt).
InWEnt (2006b) *Didaktik Leitfaden* (Bonn: InWEnt).
Kuehl, S. (2004) Fashions in development cooperation. Capacity building and capacity development as new models for development aid organizations. Munich (Unpublished).
OECD-DAC GOVNET (2006) *The Challenge of Capacity Development: Working towards Good Practice* (Paris: Development Co-operation Directorate).
Varis, O., Rahaman, M. M. & Stucki, V. (2008) The rocky road from integrated plans to implementation: lessons learned from Mekong and Senegal River Basins, *International Journal of Water Resources Development*, 24(1), pp. 103–121.

Challenges for Integrated Water Resources Management: How Do We Provide the Knowledge to Support Truly Integrated Thinking?

RACHAEL A. MCDONNELL

The Context

Ideas for linking our understanding of engineering and the natural science of water to the social, cultural and political context of an area have been muted for over 70 years, but the notion of IWRM became firmly entrenched in discussions on policy and water use during the last 15 years. The need to integrate has gained increasing credence as the interconnectedness of the many domains of water resources management was appreciated (Braga, 2001; Jonch-Clausen & Fugl, 2001). Interactions and feedback from the natural or human environments have compromised water management projects in many areas of the world.

The starting premise to an integrated approach is that there is a need to link the drainage basin and aquifer through to the near coastal zone and to develop an understanding of associated natural flows of water, energy, biota and chemicals. To this are added the changes resulting from engineering structures, whether for water withdraw or discharge. When human elements are included, dimensions such as health and economic well-being, hazards and vulnerability dynamics, legal and cultural rights, ownership and management structures, spiritual, investment needs and cost-recovery all make the development of understanding extremely complex. It is also important to then take onboard the important linked relationships that have many space and time scales, dimensions and dynamics.

The concept of IWRM, marking a fundamental shift away from the supply-demand balance equations solutions of the past, became mainstream after the 1977 United Water Conference in Mar del Plata and as any search on the Internet shows, it is a buzzword that is used frequently now. To integrate means to incorporate, join together or to amalgamate. In the past integration meant including the natural hydrological environment in engineering and economic driven water solutions. In more recent years this has been expanded to include other dimensions and leading proponents such as the Global Water Partnership (GWP) (2003) perceive it as a new water governance and management paradigm which if effective, could give long-term solutions to water problems. This is advocated through a move away from top-down, supply-led solutions dominated by the adoption of technology, towards a more decentralized basis with a consideration of water in its larger, more holistic context and an appreciation of local ideas and demand management. This concept is of course welcomed and embraces the principles adopted by various governments in Dublin in 1992.

If the breakdown of the IWRM definition of GWP (2003) by Braga & Lotufo (2008) is adopted, then its is acknowledged that water resources planning and management should consider multiple water uses in a river basin, it has multiple objectives including economic, social and environmental, it involves both coordination with other areas and levels of government, and with stakeholders in an open decision-making process.

As the Braga (2008) definition shows, integration takes on many dimensions and there are few involved with water that would disagree with the premises and concepts of this paradigm. However, the success of the drive towards IWRM has been questioned be some and Biswas (2005) asks the fundamental question "why it has not been possible to properly implement a concept that has been around for some two generations in the real work for macro and meso-level water project and programmes?" (p. 334). There is no doubting the challenges of putting into place the necessary political structures needed to put into play the theoretical ideas but there are also operational problems in enacting the various management instruments required.

Examples of this may be found in the papers of this issue. Many countries have tacitly met the deadlines of 2005 of developing plans for IWRM following the Johannesburg World Summit on Sustainable Development in 2005. Some have also embodied the tenets of the paradigm into their legal instruments, but the actual implementation of the various aspects of it to support the day-to-day water management in most countries is a long way off. This paper will consider the reasons behind one particular challenge to this, developing methods and systems that can support the information required for integrated decision making. It is important to acknowledge and address such practical considerations to ensure IWRM is not just a conceptual and academic exercise, or a ticked box on the way to securing funding for a project or programme. To begin with it is helpful to consider the nature of information requirements to support IWRM and then outline how these are currently being met and the challenges to be faced.

The Role of Information in IWRM

For informed decisions to be made in IWRM, both reliable and timely information must be available for all the aspects for the base area (river basin or aquifer discharge zone) whether control is decentralized and/or involves national water decision organizations. This is obviously fundamental to any good governance objectives to ensure that balanced, efficient and equitable decisions are made. Information, whether in the form of quantitative measured data values, textual or verbal local wisdom (GWP, 2003) or analyzed or modelled results, is used at different stages of the IWRM process and in different forms by the various authorities and stakeholders. To date there has been an emphasis in many countries on information to be managed at the river basin or discharge zone level.

The nature of the information needs will vary over time and between the levels of governance and management, which may be characterized as operational, associative or strategic (as illustrated in Figure 1 after Garcia, 2008). Initially it is important to establish and use current and historical data to characterize the baseline conditions of the area, so gaining an understanding of the state and dynamics of the various aspects of the environments For many areas this stage involves developing new monitoring networks and establishing some type of information system, usually involving databases linked to a GIS, to store and manage the data. Against this understanding of the area, problem solving, developing priorities, defining management options, and establishing decision criteria may be tackled (GWP, 2003).

At the operational level, data are needed on the day-to-day levels and status of water bodies, such as flows, quality, abstraction and discharge levels. Depending on the size of the area, there may be a breakdown into management units that are sub-basin or there may be a focus on areas of special significance. Often of interest are maxima or minima levels and their relation to regulatory limits. Other information required includes status of engineering structures, ecological needs of flow, quality at ecologically significant parts of the area, and calendars of local cultural events which all support the management needs at this level.

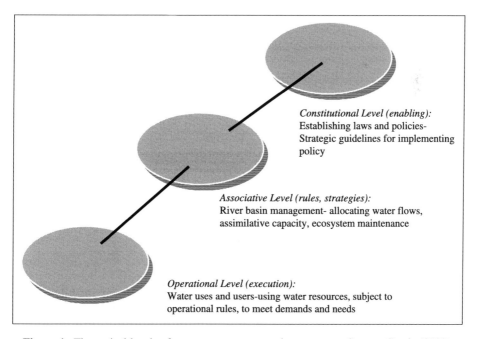

Constitutional Level (enabling):
Establishing laws and policies-
Strategic guidelines for implementing
policy

Associative Level (rules, strategies):
River basin management- allocating water flows,
assimilative capacity, ecosystem maintenance

Operational Level (execution):
Water uses and users-using water resources, subject to
operational rules, to meet demands and needs

Figure 1. The typical levels of water management and governance. *Source:* Garcia (2008).

At the associative level, information tends to be used in a synthesized form at the basin or regional scales, particularly where there may be water transfers between neighbouring areas. At this level, many forms of additional information are required such as patterns of demand, economic development indicators, ecosystem and human health. The information will also be used by those working outside the direct water management area. For example, organizations responsible for developing regional economic aims will use the base water, economic, social and environmental data to support decision making for policy and programme designs and implementation.

At the constitutional level there is a need for information which details the status of the water systems, but that also supports projection simulations of future scenarios to ensure informed decisions are made on developing and allocating resources, managing demand or processing capabilities. The information will be used by many different government agencies such as those responsible for economic development, primary industry and the environment, as well as social and welfare bodies. Information is also required by any regulatory bodies to ensure compliance with national and international objectives on water flow, quality and even ecological status.

Away from the formal management of the water and sewage provision, information is also required by other organizations such as representatives involved with any participatory organizations and NGO's. These are involved with influencing policy at many different levels so require many different forms of data.

This complex web of organizations, management and policy developments, and the role of information within, is well illustrated in the schematic (Figure 2) developed for Sao Francisco river in Brazil (Braga, 2008). This shows the types and role of information necessary to support the many aspects of water management in the river basin.

From the preceding discussion, it becomes obvious that there are a number of complexities involved with providing information to support the decision making of these various user groups:

- The information has to be available to numerous users who have variable skills and knowledge bases, and are from different disciplinary backgrounds.
- The users have different information needs, requiring data at different time and space scales, and various degrees of prior synthesis and analysis.
- The users are often geographically dispersed.
- Where public participation is active, the information needs to be available and accessible to the non-specialist.

Digital databases and Internet/Intranet based technology have important roles to play in meeting some of these challenges. Front-end systems are being developed which are easy-to-use so that users who are technically inexperienced may still access the information that is stored on centrally located and managed databases. However, in many countries, this data provision is just not possible without major investment in IT infrastructure (and training), particularly in more remote parts.

Developing Data to Support IWRM

Given the important role of information in IWRM, a prerequisite to supporting this is the provision of basic data, collected over space and time, that allows an understanding of the environmental, social, cultural and economics dynamics of an area to be developed.

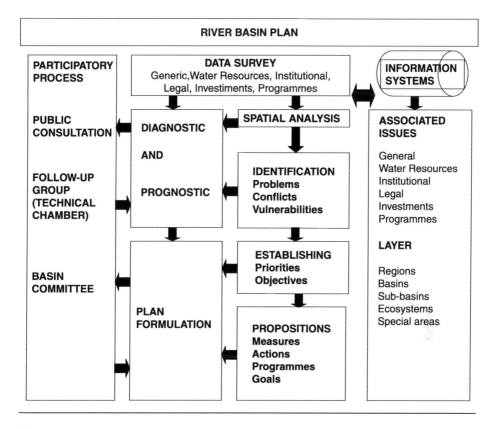

Figure 2. An example of the stakeholder interactions and information structure involved in river basin planning. *Source:* Braga (2008).

In many countries the provision of basic hydrologic and hydraulic data for major rivers and groundwater systems is adequate to begin to develop an understanding of the flow dynamics that form the foundation stone of subsequent analysis. The same can rarely be said of the tributary systems, and given that it is often on these that engineering structures such as dams are built, this brings difficulties to water resource planners. Similarly, characterization of the physical, chemical and biological elements of the water quality is usually limited by irregular or infrequent spot sampling of the river or aquifer. Whilst in recent years innovations such as automatic gauging stations and telemetry have made the collection of these data easier, the economic cost has ensured a restricted uptake so far of this enabling technology.

Away from the measurement of actual water variables, the demands for a more holistic inclusion of phenomena ensures that secondary data such as soil and land-use maps, agricultural and population censuses, ecosystem, land/property ownership and water rights, cultural and social surveys are needed to support decision making. Information technology, especially based on the Internet, has begun to ease some of the data bottleneck problems associated with using data held by other organizations, and in many countries today government and other data collection agencies are making their information available through web-based systems. The provision of meta-data alongside this has ensured that new users are aware of how and when the data sets were collected.

Other technologies such as remote sensing may be used to support rapid current acquisition of data such as topography, terrestrial ecosystem status, soil and land use, and geological structures (especially useful for groundwater systems). However, measuring other phenomena is less straightforward or just not possible. For example, whilst it is possible to gain a rough assessment of current population levels by multiplying the number of accommodation units by the average number of people per home type, this does not give important information to water managers on the types of social and economic structures within the population.

There are particular problems when defining socio-economic variables that cannot be measured as a single point such as percentage of infant mortality. These data need to be defined in terms of an area, which involves establishing a boundary that is meaningful but does not actually exist. Problems result from the selection of these artificial boundaries as different values will be recorded for a variable depending on the size and zoning of the boundaries. For example, employment rates will vary depending on whether the census tract data or its enumeration area is used. Whilst there is considerable literature on this, known as the Modifiable Areal Unit Problem (Openshaw, 1984), techniques for tackling the differences in values that result (and so affect decision making) are onboard but are still underdeveloped yet.

Using secondary data brings its own problems and these have rarely been addressed in the water literature. When data are collected it is often for a particular purpose and so is based within a specific *ontological* framework. Problems arise when these data are used for a different purpose, such as in IWRM. The variables measured and the scale of data collection adopted mean that the information required from the data is not always provided. For example, a soil survey might give spatial measures of variables such as carbonates or pH, but important values for water providers such as nitrate and phosphate levels are not sampled. Examples of this mismatch in data are often even more pronounced in the social surveys where the variables are more complex and difficult to define. Variables capturing social, structures, spiritual and aesthetic values or others linked to perceptions and senses associated with water may be of great importance to the decision-making process in some areas, but collecting them is difficult and not undertaken as part of other surveys. This has traditionally led to their omission from many analyses.

Mismatches are also often found with both the time and space scale of data available. For example, land-use maps may be available but they are at a scale of 1:500 000 and were last updated 10 years ago. Inappropriate scales of information can greatly affect our ability to gain an understanding for an area and various manipulations of the data such as basic scaling or geostatistical operations need to transform them to an appropriate scale for integration with other information. Using these methods there is an underlying assumption that the same patterns and forms are found at the different scales, which of course is often not upheld. Particular problems are found with data in which defined areas are categorized such as the number of species present or infant mortality rates. The scale of the spatial unit may be too fine or too coarse for the analysis and aggregation (summing or averaging) or disaggregation methods (e.g. genetic algorithms) must be used. Of course, these bring questions of how accurate and representative the data are.

Perhaps one of the biggest limitations to truly integrated IWRM is the poor availability of socio-economic data. Whilst population census data give an overview of the certain measures such as the age structure, religious groups and number of occupants per household which are important variables affecting water usage, information on political and social structures, gender issues other than the ratio of men to women, local forces on economic development, human and technological resources are just not regularly surveyed. Yet these variables are important for decision makers to consider under the umbrella of good governance. It has been the lack of understanding of these social variables that has undermined water projects and management in many areas. Most of the available data are collected through qualitative surveys and whilst coding the information into broad categories is useful, it is often difficult to apply boundaries to these. This obviously restricts the use of such information in a database system dependent on some form of spatial representation such as a GIS (discussed in more detail later) and its combination with highly numerical values for natural scientific data.

This section has highlighted a number of challenges in terms of data provision that the demands of the IWRM approach brings to the fore. Of course, there is a need for pragmatism and those developing IWRM need to work with what is available now and ensure both funding and innovations in technologies are used to the maxima to ease the gaps in the data. However, in the meantime it is important that resulting limitations to the understanding of a particular dimension to the analytical methods that may be adopted, and in the degree of accuracy and representation of information, are acknowledged in the development of policy and management options.

Managing the Information

Data alone cannot supply all the information required to support IWRM at the various levels of governance. Analysis involves bringing together the disparate datasets to consider the impacts, interactions and broader context of phenomena. In order to help understand and interpret the dynamics, patterns and trends in the various datasets, statistical analyses and mapping have been used for quantitative information, whilst qualitative data maybe synthesized using mapping again and textual and narrative analysis.

The technology that is most used to integrate the various datasets are GIS, which are available today as PC-based software. There are two main types of system that are based on the data model used to represent geographical phenomena (Burrough & McDonnell, 1998). Vector based systems record geographical phenomena as a series of points, lines and polygons, just as we see on traditional maps, while raster based systems use classified grid squares to show the phenomena in an area.

The development of an integrated spatial database involves first defining in the GIS a particular geographical referencing framework base (in terms of data, map projection, coordinate system). The various thematic maps, tables, etc. of the data are then input to the system as a series of layers. For each point in time a new layer needs to be established to detail the data. Obviously geo-referencing needs to be added to data where it is absent, so that it maybe integrated with other layers in a GIS. For data that are already spatially defined, their geo-referencing system must be converted to the base framework, and whilst most GIS provide a wide range of transfer functions to support this, local data often used prior to the adoption of international standards such as WGS 84 cannot always be translated (Adams, 2004).

Some of the challenges to using GIS as an integrator in IWRM are obvious socio-economic ones such as access to this type technology/data and the skills required for operation, but there are also more fundamental problems. There are a number of variables that cannot and should not be characterized as some precise bounded measure, in terms of either the raster or vector data models. It might be because it is inappropriate given their nature, such as the levels of uncertainty or because they are temporally dynamic. Stakeholders are mobile agents having influences that are not necessarily restricted in terms of geographic distance or space, for example. Networks of power and control cannot be represented within the spatial framework of a GIS. Cultural and aesthetic values are particularly difficult to characterize and it is understandable that the adoption of the GIS technology in the social sciences has been relatively limited. This has meant that the variables which are difficult to define and represent are usually omitted from analyses using this technology.

Deriving Information and Knowledge from Data

Whilst GIS allow an integrated approach to visualization and basic analysis of geographical data, more complex methods are needed to understand the feedback, interactions and dynamics of water resources systems. In the literature, analytical approaches which are based on a systems analysis approach are advocated and for the various subsystems, analytical methods such as demand assessment, Environmental impact assessment (EIA) and strategic EIA, Social Impact Assessment, risk or vulnerability assessment and simulation modelling (GWP, 2003; Bouma et al., 2005) have been used. The nature of the subsystem will dictate the scale and the structuring of any integration between methods used in the different fields of interest. For example, to analyze the impacts of a proposed water sanitation project, separate assessments tend to be undertaken on the impact on the environment, impact on society and then modelled changes to water flow and quality. Whilst there are separate critiques of these various methods (Wynne & Mayer, 1993; Cashmore, 2004), they still continue to be used and are often legally demanded in water resources development work under national legislation or donor conditions for funding.

As part of many predictive assessments, simulation modelling has been used to investigate various subsystems, especially for the natural environment, and there is a growing body of literature that links water resources models to water quality, ecological and climate variables to derive impacts on the various parts of the environment (McDonnell, 2000; Manoli et al., 2001; Cai et al., 2003; Facchi et al., 2004). In some work the groundwater and surface water systems have been integrated (Hattermann et al., 2004). These models, are often linked to GIS, use their data storage and display capabilities so that model results may be shown spatially. Some modelling systems have focused on the sectoral use of water such as agriculture and linked river simulation to agricultural planning, and hydrological modelling such as the Nile-Decision Support Tool (see www.fao.org/docrep/007/y5716b/y5716b01.htm). These developments in more complex, integrated modelling have been supported by the availability of more interactive and user-friendly modular software environments such as Stella that require less knowledge of formal programming languages, therefore supporting the water specialist in representing the system under consideration (McDonnell, 2000; Villa, 2001).

As these examples show, most the modelling developments over the last decade have focused on bringing together factors and variables in the natural environment. The ecological response has been included in a number of models (Janssen et al., 2005;

Schluter *et al.*, 2005), but one of the limits to further development is the limit to our current ability to predict responses in the biota to changes in the hydrosphere. Some simulation modelling efforts have included the economics of water used and ecology, however, there has been little inclusion in the modelling about the impact of social and cultural aspects of water management strategies. In many ways this reflects the move in social sciences away from the quantitative methods of the 1960s and 1970s towards different theory building and analysis based on more abstract representations of space and time. These new conceptual frameworks and their claims to knowledge building through changing discourses and practices have used concepts such as networks in explaining processes. This is obviously different from those used in the natural sciences (Pickles, 1999).

Integrating Analyses

Under the paradigm of IWRM, the outputs of these various analyses need to be combined to give an overall understanding of effects of various water management strategies. In the development of a number of different spatial decision support systems for IWRM, MCA (Multi-Criteria Analysis) have been used to manage in an objective and consistent way, large sets of complex information that are measured on many different metric systems. The method takes objectives, criteria for selection and weightings for managing a river basin (which are defined by the decision-making team) and then through weighted and scored matrices, they rationally assess the extent to which these objectives may be fulfilled. These all require the conversion of data to some standardized quantitative measure to allow some type of weighted analysis and comparison across the subsystem boundaries. The MULINO Project is a good example this type of work (see www.siti.feem.it/mulino/).

The integrating methods of indicators and indices have also been used and these give a useful synopsis of variables such as economic return on water used, number of species per area, etc. (GWP, 2003). More developed indices such as the Water Poverty Index (Sullivan, 2002; Sullivan *et al.*, in press) and the more detailed Climate Vulnerability Index (Sullivan & Meigh, 2005) provide a better understanding of the relationship between the physical availability of water, its ease of abstraction, and the level of welfare, and they are used integrate various data to define five main components (resources, access, capacity, use and environment). From this it is possible to synthesize and categorize the water resource situation.

These various methods are by design synthesizing techniques that support the integration of many forms of data to a series of single values or matrices from which it is possible to derive rationalized and objective preferences. However, the extent to which they provide useful *knowledge* to decision makers is open to debate. Both MCA and the indicator/index approaches are based on rational, deductive ideals in which empiricism is used to bring an objective basis to conceptualize an environment or derive preferences for a particular set of actions. MCA may be criticized for providing rather simplistic choice models based on average values which can only be used for rationalizing variables and not for predicting or developing causal linkages between them. The complexity and the various interactions and feedbacks between the variables, which are often the unplanned and sometimes bring unwelcome side effects of a management strategy, are not represented.

The limited acknowledgement of the influence of scale in these methods is also of concern. Many are undertaken at one particular scale, often dictated by the data or by the management units involved. This will not always be optimum to developing an understanding of natural and social science phenomena, especially where scaling or aggregating/disaggregating operations are needed to transform available data. The resulting accuracy limits to the outcome of analyses or modelling need to be acknowledged by those using the information.

Administrative and Political Challenges to Developing Information Support Systems

So far the discussion has focused on the development of data systems and analytical and assessment methods to support decision making in an IWRM framework. Information systems are needed to manage and share, and their design and structure should be developed following extensive discussions with potential data providers and users to ensure it will meet future as well as present requirements. Discussions will also help engender a communal sense of ownership which will help to maintain the system after the initial establishment stages are completed.

Such developments of information systems are not undertaken in isolation from the political environment. The provision of information, whilst minor in comparison to the political challenges of developing equitable, efficient and safe water and sanitation services, brings with it a number of problems that have in some areas stymied the ability for fully integrated decision making. Underlying many of the challenges is the fundamental tenet that data gives power and its collection and management is a financial cost to any department or organization whatever the level of government. There is often consequently a reluctance to share data with other (rival) departments. With the need in IWRM for substantial secondary data, collected by organizations not directly related to water management, this means that there are administrative and political questions that must be addressed before common stewardship of a data system may develop:

- Who will own and who will manage the Information Systems?
- Who are the data providers? Who needs the data?
- What structures need to be in place to support data sharing?
- What standards will apply to the data?
- Who will pay for it? What is the time scale?
- Who is legally responsible for maintaining its accuracy and currency?

The development of information systems requires a commitment to long-term funding that extends well beyond the development stages of an initial funded project. This is particularly true when this funding comes from donor agencies.

Social and Ethical Issues of Developing Information Systems to Support IWRM

Given the importance of GIS and associated databases to developing an information infrastructure, it is important to consider the broader impacts such a move would make. Introducing information systems into any society means bringing a new series of institutions, discourses and practices into play (Pickles, 1995, 1999). The increased ease in accessing and using data and developing information brings many benefits, with the ultimate being decision making based on a wide and integrated knowledge of an area.

The ease of access to organizations and stakeholders not directly involved with the day-to-day management of water will also ensure water issues are taken on board in decision making in other sectors and levels of government other than the immediate basin or discharge zone. There are also ideas that the role of civil society will be enhanced with data provision, leading to informed and empowered public participation.

There is obviously a converse side to this, and it is important to consider the social structures of a particular setting and the impact information systems will have on the various strata of civil society and the roles they may then take in the decision-making process. For some groups, often the most vulnerable, the lack of access to this data and fear of technology can marginalize them from the decision-making process. The differential access will ensure the information is not available to all equally and there have been various critiques of the politics of knowledge (Curry, 1995; Pickles, 1995).

True Integration in IWRM: Possibilities and Challenges

The formal acknowledgement of the need to include and integrate into decision making the various sectors, governance structures, people and environments involved with and influenced by water is a major step forward to realizing efficient, equitable and sustainable water management. Of course, the availability of knowledge is one of the foundation stones to support this. However, there are problems and challenges with some of the approaches that have been put forward. The very practical problems, such as availability of data measuring the appropriate variable and at a suitable time and space scale, are of course real. The decisions made by water managers need to be based on information for which some notion of reliability and accuracy are known. From the vast literature in the field of GISscience, it is known that there are many conceptual problems such as dealing with uncertainty in the data, data accuracy, error propagation, acknowledging the impacts of scale of data and process representation that have not been addressed in the current discussions on methods and information for IWRM.

However, it may be argued that the biggest controversy is away from the basics of data provision and lies with the positivist, empiricist and technocratic approaches to analysis and information development. The notion that a series of layers of spatial data, linked through subsystem assessments and mathematical modelling and combined using weighting and matrix-based procedures, can give suitable *knowledge* of the complexities of environmental, economic and socio-cultural and political interactions, has not been substantiated through successful applications. It could be argued that the information systems that are required to manage the data, also have the undesired impact of affecting both the development of knowledge and the data that may be used.

The result has been a rather techno-scientific set of approaches with the greatest weaknesses being in describing, analyzing and developing understanding of the influences of water development ideas by, and on, the many structures of a society. In many ways the methods for developing information ignore the last two decades of work in social theory. In these there is an abstracted view of the geographical space and the influences and interactions are often defined in terms of networks and flows of power between the various actors/stakeholders involved with governance. Social surveys use other methods to characterize the various groups and impacts within that again do not fit well within this positivist setting.

At present the possibilities for truly integrated water resources management are limited, not by a conceptual framework, but by the ability to really represent the full dimensions of variables, interactions and complexity that come into play in any water management project or policy. There has been a move to use off-the-shelf existing methods, but this new conceptual framework needs new methods. There is a need to work with research groups, such as those in the Geographic Information Science, ecologist and social sciences to develop new methodological approaches to support the important ambitions of IWRM. These methodological challenges are being addressed in many other areas of natural resource management such as in forestry.

There is also a need to question whether a single paradigm of IWRM can be translated to all environments given the complexities in natural, social, political and economic phenomena. It is without question a desirable framework for water management, but it is not hard to see why it is just not possible for all countries, so there should be parallel moves to develop other ideas which bring the same returns of equity, efficiency and sustainability.

References

Adams, R. (2004) Seamless data and vertical datums—reconciling chart datum with a global reference frame, *The Hydrographic Journals*, 113, pp. 9–14.

Biswas, A. K. (2005) Integrated water resources management: a reassessment, in: A. K. Biswas, O. Varis & C. Tortajada (Eds) *Integrated Water Resources Management in South and South-East Asia*, pp. 319–336 (Oxford: Oxford University Press).

Bouma, J. J., Francois, D. & Troch, P. (2005) Risk assessment and water management, *Environmental Modelling and Software*, 20, pp. 141–151.

Braga, B. P. F. (2001) Integrated urban water resources management: a challenge into the 21st century, *International Journal of Water Resources Development*, 17, pp. 581–599.

Braga, B. P. F. & Lotufo, J. G. (2008) Integrated river basin plan in practice: the Sao Francisco river basin, *International Journal of Water Resources Development*, 24(1), pp. 37–60.

Burrough, P. A. & McDonnell, R. A. (1998) *Principles of Geographical Information Systems* (Oxford: Oxford University Press).

Cai, X., McKinney, D. C. & Rosegrant, M. W. (2003) Sustainability analysis for irrigation water management in the Aral Sea region, *Agricultural Systems*, 76, pp. 1043–1066.

Cashmore, M. (2004) The role of science in environmental impact assessment: process and procedure versus purpose in the development of theory, *Environmental Impact Assessment Review*, 24, pp. 403–426.

Curry, M. R. (1995) Rethinking rights and responsibilities in GIS: beyond the power of imagery, *Cartography and Geographic Information Systems*, 22, pp. 58–69.

Facchi, A., Ortuani, B., Maggi, D. & Gandolfi, C. (2004) Coupled SVAT-groundwater model for water resources simulation in irrigated alluvial plains, *Environmental Modelling and Software*, 19, pp. 1053–1063.

Garcia, L. (2008) Integrated water resources management—a 'small' step for conceptualists, a giant step for practitioners, *International Journal of Water Resources Development*, 24(1), pp. 23–36.

GWP (Global Water Partnership) (2003) *Integrated Water Resources Management Toolbox: Sharing Knowledge for Equitable, Efficient and Sustainable Water Resources Management* (Stockholm: Global Water Partnership).

Hatterman, F., Krysanova, V., Wechsung, F. & Wattenbach, M. (2004) Integrating groundwater dynamics in regional hydrological modelling, *Environmental Modelling and Software*, 19, pp. 1039–1051.

Janssen, R., Goosen, H., Verhoeven, M. L., Verhoeven, J. T. A., Omtzigt, A. Q. A. & Maltby, E. (2005) Decision support for integrated wetland management, *Environmental Modelling and Software*, 20, pp. 215–229.

Jonch-Clausen, T. & Fugl, J. (2001) Firming up the conceptual basis of integrated water resources management, *International Journal of Water Resources Development*, 17, pp. 501–510.

Manoli, E., Arampatzis, G., Pissias, E., Xenos, D. & Assimacopoulos, D. (2001) Water demand and supply analysis using a spatial decision support system, *Global Nest: The International Journal*, 3, pp. 199–209.

McDonnell, R. A. (2000) GIS-based hierarchical modelling of the environmental impacts of river impoundment, *Hydrological Processes*, 14, pp. 2123–2142.

Openshaw, S. J. (1984) *The Modifiable Areal Unit Problem* (Norwich: Geobooks).

Pickles, J. (Ed.) (1995) *Ground Truth: The Social Implications of Geographic Information Systems* (New York: Guildford Press).

Pickles, J. (1999) Arguments, debates and dialogues: the GIS-social theory debate and the concern for alternatives, in: P. A. Longley, M. F. Goodchild, D. J. Maguire & D. W. Rhind (Eds) *Geographical Information Systems: Principles and Technical Issues*, pp. 49–60 (New York: John Wiley).

Schluter, M., Savitsky, A. G., McKinney, D. C. & Lieth, H. (2005) Optimizing long-term water allocation in the Amudarya River delta: a water assessment model for ecological impact assessment, *Environmental Modelling and Software*, 20, pp. 529–545.

Sullivan, C. A. (2002) Calculating a water poverty index, *World Development*, 30, pp. 1195–1211.

Sullivan, C. A. & Meigh, J. R. (2005) Targeting attention on local vulnerabilities using a integrated index approach: the example of the Climate Vulnerability Index, *Water Science and Technology*, 51, pp. 69–78.

Sullivan, C. A., Lawrence, P. & Meigh, J. R. (in press) Application of the Water Poverty Index at different scales—a cautionary tale. *Agriculture Ecosystems and Environment*

Villa, F. (2001) Integrating modelling architecture: a declarative framework for multi-paradigm, multi-scale ecological modelling, *Ecological Modelling*, 137, pp. 23–42.

Wynne, B. & Mayer, S. (1993) How science fails the environment, *New Scientist*, 1876, 5 June pp. 33–35.

Index

Page numbers in **Bold** represent Tables and Figures

For Product Safety Concerns and Information please contact our EU representative GPSR@taylorandfrancis.com Taylor & Francis Verlag GmbH, Kaufingerstraße 24, 80331 München, Germany

Printed and bound by CPI Group (UK) Ltd, Croydon, CR0 4YY

02/05/2025

01859533-0001